LABORATORY ACCREDITATION AND DATA CERTIFICATION

A System for Success

CARLA H. DEMPSEY
J.D. PETTY

 LEWIS PUBLISHERS

Library of Congress Cataloging-in-Publication Data

Catalog record is available from the Library of Congress.

ISBN 0-87371-291-9

LEWIS PUBLISHERS, INC.
121 South Main Street, P.O Drawer 519, Chelsea, Michigan 48118

PRINTED IN THE UNITED STATES OF AMERICA

THE AUTHORS

Carla Dempsey is a Program Development Manager for Environmental Programs at Lockheed Engineering and Sciences Company in Washington D.C. She is responsible for locating opportunities in the environmental field for applying the high technology that has been developed at Lockheed.

Ms. Dempsey has prior experience at the Environmental Protection Agency where she was the Quality Assurance Coordinator for the Superfund Contract Laboratory Program. She has also been a Project Manager with environmental consulting firms. Before focusing on environmental issues, she assisted the U.S. Navy design a national quality assurance program as the Quality Assurance Manager for the Laboratory Division at Puget Sound Naval Shipyard in Bremerton, WA.

Ms. Dempsey received her BS. in Chemistry from Gonzaga University in Spokane, WA and her MS in Chemistry from the University of Washington in Seattle.

J. D. Petty is the Director of Analytical Research and Quality Control at the Fermenta Animal Health Co. The research involves development of methodology to determine a wide array of pharmaceutical ingredients in a variety of matricies. In addition Dr. Petty is responsible for ensuring that all analytical data meet the requirements of FDA Good Laboratory Practice and Good Manufacturing Practice regulations.

Dr. Petty has prior experience as the Chief of the Quality Assurance Research Branch of the EPA's Environimal Monitoring Systems Laboratory at Las Vegas and as the Chief Chemist of the U.S. Fish and Wildlife Services' National Fisheries Contaminant Research Center located in Columbia, Missouri.

Dr. Petty received his BS in Chemistry from Central Missouri State University in 1970 and his Ph.D. in Organic Chemistry from the University of Missouri-Columbia in 1975. Dr. Petty has authored and co-authored over 50 publications on a wide range of research topics.

PREFACE

Critical decisions are often based on chemical data. Therefore, it is vitally important that the quality of the data that are gathered be sufficient to allow good decisions to be made. Most people believe that the quality of laboratory data available in this country does not directly affect them. This is not true. Literally thousands of decisions that are based on laboratory data are made each day.

National decisions that affect the entire American public are made using environmental data. The government routinely uses data to determine which hazardous material disposal sites are ranked as Superfund sites. The sites that are identified as the most hazardous are scheduled for clean-up actions, since they are considered the most harmful to human health and the ecology.

How do decision makers determine if the data upon which decisions are based is valid or invalid? Many professional and trade organizations have attempted for years to assist the public in identifying entities that are capable of generating quality data. Accreditation and certification programs are two ways that have been used in the past to differentiate between acceptable and unacceptable laboratories. The current systems, however, have been ineffective in assisting data purchasers determine which laboratories can generate quality data for specific purposes.

It is the intent of the authors to provide a new perspective to an old problem of how to differentiate between acceptable and unacceptable data. In order to establish a basis for the in-depth presentation of the problem in this book, the authors examine the problem as it is perceived by users of data, the government, the laboratory industry, and certifying the accrediting bodies.

The authors describe the usual ways that have been employed in the past by accreditation and certification bodies to judge the competence of the testing laboratories that supply data. The generic requirements of accreditation, certification, and related programs are briefly outlined. Conclusions about the merits and weaknesses of the current approaches are drawn. Finally, recommendations for a practical approach to establish competence of testing laboratories is provided.

This book is different from any other books written about this subject because the authors have grappled with the problem at the hands-on level and from many different perspectives. The authors have extensive laboratory experience in a myriad of different types of laboratories, from metals testing to medical laboratories and also have experience in the design, improvement, and implementation of national laboratory evaluation programs. The authors also have experience in both the public and the private sector.

A working system that integrates many of the concepts from existing accreditation, certification, and registration systems is described. The basic system concepts are designed to facilitate its use as the basis of a national accreditation and certification system that incorporates third-party involvement. The environmental laboratory industry has been used as the template for the model because of the author's common experiences in the industry. It was also chosen because the System for Success was based, in part, on a current environmental laboratory and data assessment program. Even though the system is designed primarily with the environmental laboratory industry as the focus, the concepts presented in this book are applicable to other types of laboratories and are valid in any other laboratory accreditation area.

In addition to presenting a strategy to improve the current laboratory accreditation and certification issue, this book endeavors to inform all parties about the responsibilities of data users and producers in generating and acquiring appropriate data. All data are not created equal. Data purchasers must specify what type and quality of data are needed and must assess the data that are purchased to determine if it meets their specifications. Laboratories must implement internal quality assurance programs that will assure data purchasers that the data they purchase will meet their expectations.

Education of the data users in their responsibility to define the criteria for inspection of data is a critical key to an effective testing laboratory accreditation and certification program. Data users must not purchase data by price alone. Data are generated for its information content. People use information inherent in data as the basis for making decisions. Therefore, all data users have the responsibility to acquire the appropriate quality of data needed for decisions. A good accreditation and certification program can assist them in choosing laboratory services. However, accreditations and certifications will never be the entire solution. They are only tools to assist in choosing a laboratory and in monitoring the laboratory's performance.

Finally, this book attempts to spearhead a movement to assess and eliminate wherever possible the conflicting and redundant requirements among the many certification, accreditation, registration, and other related systems that are in effect in this country. The effect of the unnecessary duplication of effort is costly to the consumer and the laboratory industry alike. Since the consumer for the majority of environmental and regulatory required data is the taxpayer, the effect of reducing the redundancy and waste will allow better use of taxpayer's funds.

In addition, perhaps the most important reason to eliminate the redundant efforts and develop a credible cost-effective program is because the result of redundant, politically motivated, and unnecessary accreditation and certification programs is higher cost to data purchasers. Even though all of the programs were developed and instituted to assure the quality of testing laboratory data, they waste QA resources and decrease American productivity. This country must expend each dollar for QA wisely in order to attain a global competitive edge. The implementation of redundant QA programs is damaging to the purpose of QA and may ultimately lead to poorer quality data than if the systems were not in place. The old saying "When quality is everybody's responsibility — it becomes nobody's responsibility" is appropriate in describing what happens when redundant accreditations are implemented. If there are many redundant systems, there is a tendency to hope that some other QA system will find what your system does not. This leads to quality becoming "nobody's responsibility" and lower overall quality.

Once a laboratory gets the required seal of accreditation approval, is it realistic to assume that it implements the costly, redundant, and conflicting requirements of all the competing accreditation systems in its operations? Unless the accrediting body inspects the facility or its data continually, it is naive to assume that the laboratory meets all the accreditation systems' redundant and conflicting requirements on a day-to-day basis.

What then, is the value of the accreditation or certification "seal of approval"? Does it allow the purchaser to believe that the laboratory is producing acceptable results? It does allow the laboratory to post the "seal of approval" so that other purchasers do not feel the need to inspect data because of the laboratory's accredited status. If the accreditation removes this skepticism, is it not realistic to assume that the data generated on a day-to-day basis would, at best, remain the same quality? At worst, the data could deteriorate without continual scrutiny.

The authors have presented the current situation and offer a practical solution to the problem of laboratory accreditation. It is hoped that all those who are interested in improving the current situation can focus on this solution and use it as the basis for a collective solution. It is also hoped that those who are interested in developing a national accreditation and certification program realize that the system must be rigorous and provide continuous monitoring. If anything less is developed, it will have no value for data purchasers, and will likely promote the production of lower quality data. The solution presented in this book is a balanced design and is a workable solution. The solution includes the critical two elements — a capability assessment and also monitoring of laboratory products to assess actual performance.

ACKNOWLEDGMENTS

"What a person thinks on his own without being stimulated by the thoughts and experiences of other people is even in the best case rather paltry and monotonous."

A. Einstein

The authors gratefully acknowledge the many people that have contributed ideas to the concepts in this book. Many chemists, government workers, engineers, and other technical specialists have increased the authors' understanding of the subject and improved their perspective. Because of the many countless discussions with these individuals, the book is not a compilation of philosophical interludes of what might work from a chemist's or an attorney's point of view. Rather, it is based on the practical considerations of many, often conflicting, views of all of these individuals.

Special thanks are given to the skeptics of a national accreditation or certification system. You individuals solidified the questions and hurdles that this system will be subjected to by critics of the national accreditation and certification concept. You took the time to question and debate many issues that may not have been seen as critical to the authors. Many of you would like to see a national program implemented, but do not think that implementation can succeed. We hope that you ask those who will be roadblocks to implementation the same type of skeptical "why?" that you asked the authors of this book. People like you are needed to work together on the solution and improve the framework that is presented in this book so that it meets your individual needs.

And last but certainly not least, the authors recognize the patience and understanding of their families and business associates for enduring the division of their attention during the writing of this book.

TABLE OF CONTENTS

LABORATORY ACCREDITATION AND CERTIFICATION — HISTORICAL PERSPECTIVE AND DEFINITION OF NEED

DEFINITION OF NEED

> The time has come the people said, to speak of many things, of dirty skies and dying ecosystems and ozone disappearing, of toxic waste and fraudulent claims and criminal profiteering ...
>
> with apologies to the Walrus

Perhaps our world is less subject to capricious whims than the world encountered by Alice, but the industrialization of society has created a situation in which our lives are based on seemingly mysterious and often incomprehensible phenomena. Few among us are capable of understanding the technology of something as mundane as compact disc players, let alone the complexities of the space shuttle, super-computers or modern medical practices. The very terminology associated with today's engineering and technology serve to stupefy even the most wide ranging of intellects.

Considering the complexity of today's technologically based society that has given us a standard of living undreamed of by even the most prophetic of futurist, how do we ransom ourselves to the ever increasing sophistication of technological advancement? Yet the situation is not bleak and foreboding. Only a few truly understand aeronautical engineering or pharmacokinetics. However, we benefit from these sciences when we fly at near supersonic speeds and are cured from illness by antibiotic therapy. One does not need to have a vast understanding of the technology in order to benefit from it. In a society that depends on technology, this is as it should be. The ransom is our belief in technology and our faith in information regarding the safety, efficacy and quality of the products of technology. This information comes from a myriad of sources and is ultimately the basis of our making use of the marvels of modern technology.

"Information is power" is an axiom often repeated from the halls of government to corporate board rooms. If we are to enjoy the abundant fruits of technology we must accept much on faith, not on technical information. We must have faith that we are being provided correct information by Government regulatory agencies, corporations, and providers of services that we are incapable of providing for ourselves. This information is more often than not based on abstract data generated by highly trained technical personnel using equipment, instrumentation and methodology incomprehensible to all but a few specialists. Is our faith in this information truly justified? Regrettably, there are numerous examples of decisions that have been made based on less than reliable data. More onerous are examples of the intentional falsification of data.

Perhaps the most notorious recent example of the generation of fraudulent data is the case of Industrial Bio-Test Laboratories (IBT). A brief review of this dark side of independent testing laboratories is instructive. This situation has been chronicled in a variety of media but the article[1] which appeared in the Amicus Journal is a germane summation. The following is a direct quote from this article and provides a good depiction of the sordid affair:

> Within the fervid, unseemly world that was Industrial Bio-Test Laboratories, the place where things turned gruesome was a room they called the "Swamp". In 1970, IBT's directors installed a Hoeltge automatic watering system for one large animal feeding room midway through Number Three building. Although it was designed to fill drinking bottles and flush wastes from hundreds of rodent cages, the equipment rarely worked properly. Faulty nozzles sprayed the room with a continuing chilly mist, showering the caged animals. Water streamed off cages and racks, submerging the floor under a four-inch deep pool. Mice regularly drowned in their feeding troughs. Rats died of exposure. No technician entered the swamp without rubber boots, and many wore masks to protect themselves from the hideous stench of disease and death.
>
> During the course of a two-year feeding study, involving more than 200 animals, the mortality rate in the swamp reached 80 percent. Worst of all was cleaning the cages. Dead rats and mice, technicians later told federal

investigators, decomposed so rapidly in the swamp that their bodies oozed through wire cage bottoms and lay in purple puddles on the dropping trays.

It was in conditions like these in the Swamp and four other major animal feeding areas that IBT conducted thousands of critical research projects for nearly every major American chemical and drug manufacturer, dozens of foreign concerns, and several federal agencies as well. Nearly half of IBT's studies were used to support federal registrations of a mammoth array of products: insecticides, herbicides, food additives, chemicals for water treatment, cosmetics, pharmaceuticals, soaps and bleaches, even coloring for ice cream.

One of the nation's oldest independent laboratories, during its last decade, IBT was also the largest, performing more than *1,500 studies* in its main facility in Northbrook, Illinois, twenty-five miles north of Chicago, and in two satellite laboratories in Neillsville, Wisconsin, and Decatur, Illinois. It has been estimated that between 35 *and 40 percent* of all toxicology tests in the country were conducted by IBT.

Still, for all its prosperity and spurious prestige, IBT's business crumbled rapidly starting in 1976, when at the zenith of the lab's corporate strength, investigators from the U.S. Food and Drug Administration (FDA) uncovered what they allege is the most massive scientific fraud ever committed in the United States, and perhaps the world.

Examples of falsification of data were found by investigators to be widespread in many IBT studies. Why then was IBT so successful as a toxicology laboratory? Simply put, "IBT became the largest testing lab in the country because companies knew this was the place to get the results they wanted." The exact reasons why IBT fell victim to the demon of falsifying data might never be known — perhaps greed, mismanagement, pressure to "get the answer" or a combination of all these factors plus others that are more subtle. What is demonstrated and indisputable is this — the products tested by IBT became a part of our society. Following the revelation of data falsification, the FDA and Environmental Protection Agency (EPA) required companies to re-run many of the tests conducted by IBT. Fortunately, no serious health problems for the American public resulted from this deplorable situation. However, the cost of re-testing was undoubtedly passed on to the public.

A positive result of this reprehensible affair was the codification of the FDA's Good Laboratory Practices (GLP).[2] These criteria are an attempt to ensure that all laboratories produce analytical data that meets minimum standards. The FDA now periodically evaluates laboratories that generate data in support of their program for adherence to these standards. A wide array of programs that are designed to ensure data quality have grown from this first GLP program. Nearly all government regulatory agencies have a GLP based program. The response of government regulatory agencies in implementing GLP programs is positive.

While economic pressure might have been a contributing factor in the case of IBT, political pressure can be equally as detrimental to the generation of reliable analytical data. The case of the hazardous waste site commonly known as Love Canal is a prime example. Without relating all the details, it is sufficient to say that the public outcry associated with the revelation of the wide spread dumping of hazardous waste around the Love Canal in the city of Niagra Falls, New York caused a questionable political response. The problems — both perceived and real — accompanying the disposal of hazardous chemical waste certainly required answers. The answers that were needed could only be addressed by the generation of high quality analytical data for a variety of chemical residues. At the time that the hazardous waste dumping was discovered, only a limited number of laboratories were capable of performing the state-of-the-art analyses that were required to produce the correct data to answer the pressing questions surrounding the hazardous waste dumping. Unfortunately, the political pressure to obtain answers quickly determined how the data would be obtained. The result was a pronouncement that the required analyses *would* be performed in an unrealistic time frame. The results could have been, and were, predicted. The data generated were inadequate to satisfy any of the concerned parties. As a result, studies relating to the habitability of this area had to be repeated at a significant cost to the public, in terms of both dollars and loss of confidence in government agencies and technical data.

Other examples[3,4] can certainly be cited — incomplete or fraudulent toxicological data, poor quality data generated to support Resource Conservation and Recovery Act (RCRA) requirements, inferior products, shoddy manufacturing processes, pesticide residues in excess of action limits, cover-ups of illegal dumping operations, discharges of hazardous waste, and so on. All have one common thread; that thread is analytical data. The generation of data, particularly analytical data, is a multi-billion dollar per year business. How then do we both as producers of data and consumers of products and services that require data in their production, ensure that such data are of sufficient quality to satisfy its intended use? This question is of central importance in our technological society. Laboratory accreditation and data certification may be a crucial part to the answer for this question.

THE GOALS OF LABORATORYACCREDITATION/
CERTIFICATION

Let us briefly consider the plight of a decision maker who is attempting to determine if the data that is provided for use in making a critical decision are good data for that decision. The data are now gathered and presented in tables in standard report format for ease of use. But what does the person making the decision know about the quality of the data? What enables the decision maker to trust the numbers?

The assumed, and sometimes stated, goal of laboratory accreditation is to assure data users that a laboratory produces quality results. Can a decision maker rely on accreditation systems for assurance of the quality of a laboratory's results? More importantly, does laboratory accreditation assure a decision maker that the quality of the data gathered for any critical decision is the required quality?

Does the decision maker question the quality of the data? After all, the decision maker receives a report that has numbers arrayed in concise tabular form. All the numbers might appear to be in order. It is an unusual individual who understands the numerous steps and operations that are conducted to obtain the numbers that appear in the concise tables. Each operation and each step is an opportunity for error. In fact, many errors could be inherent in the numbers that are presented for the decision maker's use.

Is it the decision maker's responsibility to know how good the data are? If the decision maker is responsible for the consequences of the decision, he must be responsible for the quality of the data used to make the decision. Analytical data are often the basis for decisions that affect large numbers of people. Responsibility for acquisition and use of the correct data to make decisions rests with the decision maker.

What should the decision maker ask about the quality of the data? There are many different types of data, all produced by numerous types of analytical methodologies. Individual methods can be performed under rigorous quality control requirements for exacting uses or less stringent quality control requirements for less demanding uses. Further, the data can be produced following strict chain-of-custody procedures or can be produced with minimal written records. In addition, the data might be produced and reviewed only by a laboratory supervisor, or might be scrutinized in great detail by an independent chemist from another organization. What about these and other aspects of the data collection process does the decision maker need to know?

The intricacies of data collection are usually transparent to decision makers and are certainly unknown to the people that are affected by the decision that is made. Individual people and the general public rely on the people making decisions to collect and use appropriate data in their deliberative processes. They rely on decision makers to consider what type, quality, and quantity of data are required for each decision. These determinations should be made well in advance of the actual taking of a sample and the subsequent analysis. The considerations that are made during the planning stages determine if the numbers that will be produced to support the decision are good enough to allow a correct decision to be made. Decision makers usually rely on those individuals in their organizations that either generate data or determine if the required quality and quantity of data have been gathered. The people that procure data must have a complete understanding of the decision makers' data needs in order to correctly chose the requisite quality and quantity of data. It is the decision makers' responsibility to convey their data needs effectively in order that the required data are acquired for making decisions.

If samples can be analyzed by the people responsible for making decisions and those people have the knowledge, skills, time, and equipment to produce the required quality data, then the assessment of the quality of the data is simple for the decision makers. Usually, however, the decision makers have to delegate their responsibility for determining if purchased data are of the required quality to the people that are responsible for acquiring the data. The people that are delegated responsibility to generate the data must either acquire the samples and perform the analyses themselves or purchase the sampling and analyses. At the point of sample analysis, the data are produced according to the specifications that the decision maker communicated to the persons responsible for acquiring the data. If the specifications were incorrect or not communicated effectively, then the resulting data could be unusable for the decision making process. It is critical for the proper planning of data needs and requirements to be performed before it is possible to acquire the correct data.

It is not the laboratory's responsibility to determine what data a decision maker needs. Laboratories can assist decision makers in choosing the best means to generate the data by working cooperatively with the decision makers or their organizations, but the final choice of data, method, detection levels, and other such criteria must be made by the customers that purchase the data from laboratories.

The subject of sampling is not addressed here. The critical nature of acquiring an adequate sample before the analysis is recognized. However, it is, in itself, the topic for another volume. This book concentrates on the difficulty that data purchasers encounter in acquiring quality data from analytical laboratories.

If the people responsible for acquiring data purchase the data, rather than performing the analyses themselves, they must somehow measure "something" about the laboratory or the data to determine if the data are of the required quality. What can these people use to convince themselves and the decision makers that the data from particular laboratories are quality data? This question is the essential basis of laboratory accreditation.

The object of the above discussion about data and decision makers is included to illustrate to the reader that decision makers delegate their responsibility for determination of data quality to others. This is a great responsibility because the correctness of decisions rely on the quality of data. The key question is, "How do the people that purchase data assess the testing laboratories' abilities to perform quality work?" An additional related question is, "How do these people assess their in-house laboratories if they acquire data from in-house sources?"

In the past, the question was asked countless times and the search for a reasonable answer to the question brought about the many different accreditation, certification, registration, licensing, and inspection systems that are currently in effect. Responsible decision makers determined that they must take full responsibility for assessing the quality of the data from various testing laboratories. It can safely be stated that there are as many programs to determine competence of testing laboratories as there are managers that take their decision making responsibility seriously and can afford their own laboratory accreditation or certification systems.

A manager's own unique system might be considered the best approach by each manager if each individual is questioned about their data needs. However, the logical perception of an outside observer is that many of the systems assess laboratories in almost exactly the same way.

The most skeptical of outside observers could argue that each decision is slightly different and that each decision maker is acting prudently in creating a separate system. It could further be argued that it is nearly impossible to trust another accreditation, certification, registration, or licensing system because the quality of that system might be unknown. These observations have merit and must be addressed before a national system can be accepted.

Let us consider how prudent managers that do not have the budget or personnel to create their own accreditation systems determine laboratory capability and data quality. These unfortunate individuals are faced with the onerous task of unraveling the mysteries of the other systems that are currently in existence. Some of the questions that could be asked about the many systems are:

- What does the accreditation or certification mean?
- Is the certification or accreditation simply a license that is purchased to do business?
- Does the certification or accreditation mean that the establishment has passed certain requirements?
- If the establishment has passed certain requirements, are these requirements relevant to the type of work that the manager's need to be performed?
- Does the certification or accreditation mean that the facility was inspected? Is it a program that relies only on the laboratory's completion of an application that asks if the requirements have been met by the facility? Is the application followed by an on-site verification of the application information?
- Is the accreditation or certification performed by a trade association or is it performed by an outside party that has no stake in the outcome of the process?
- Is the registration simply a legal fee requirement that implies nothing about the quality of the establishment?
- If the laboratory is certified or accredited by one or several organizations, are the certifications or accreditations relevant to the type of analysis that the managers need to be performed?

One manager going through this process could determine from the answers to these questions that the laboratory has numerous accreditations, certifications, licenses, inspections, and registrations. Also learned from the answers to these questions is that many of the accreditation and certification systems could be strictly legal or trade association requirements that most laboratories must have in order to do business. It could also be discovered that many of the accreditations and certifications are not relevant to the type of analyses that are needed by the manager. In some cases, the manager might analyze the individual requirements of the systems carefully to determine if a laboratory is capable of providing quality data for the specific type of analyses that will be required.

It is obvious that this manager might be able to use the accreditations and certifications to determine if the equipment, personnel, facility and QA requirements might be acceptable. This is possible if the types of analyses for which the laboratory is accredited are similar to the needed analyses. If this is not the case, the accreditations are meaningless to the manager. Take, for example, accreditations of a laboratory to perform paint testing. Could this laboratory be expected to perform air monitoring for trace volatile constituents? In all probability, the laboratory would not be able to perform the analyses and would not have the equipment or facilities to perform the analyses. If however, the accreditations and certifications were in the area of waste water analysis, could the laboratory be expected to perform well on hazardous waste samples? The accreditations might indicate to the manager that at least some of the equipment for hazardous waste samples might be available at the facility. Accreditations would not be sufficient to determine if the laboratory were capable to perform quality analyses on the samples of interest, but would provide some information about the facility that could lead to further questions.

If a careful inspection of the laboratory's accreditation and certifications show that one or more of the accreditations and certifications seem relevant to the type of samples that are to be analyzed, more must be known about the requirements of those accreditation and certification systems. Such questions could include:

- What attributes does the accreditation and certification body inspect? What are the minimum requirements for these attributes?
- Is an on-site inspection performed to verify application form information? If so, what are the credentials required of the inspection officials?
- Is the establishment required to successfully perform on a performance evaluation (PE) sample? How often does the laboratory analyze a PE sample?
- Is an on-going, continual inspection and monitoring of the laboratory and the data produced by the laboratory performed by the accreditation body?

If the manager fully understands the intricacies and requirements of all relevant accreditation and certifications, this person would probably "feel" better about selection of one laboratory over another. However, this person's choice of a "good" laboratory based on accreditation and certification programs that are currently assessing laboratories could be flawed. This is because many certification and accreditation programs do not provide the in-depth assessment that the manager needs to determine whether or not the data produced by the laboratory is the quality needed. The following are existing situations that are not effectively addressed by most certification and accreditation programs.

- The laboratory could have all the facilities, equipment, personnel, and procedures available to perform quality work. Having these required attributes only provides evidence for capability — not the evidence of successful production of quality data.
- The laboratory could have all the facilities, equipment, personnel, and procedures available to perform quality work on a performance evaluation (PE) sample. This provides evidence of successful production of quality data on only one sample — not on all samples. Rarely do PE samples get treated as routine samples. They are usually assigned to the best analysts, are analyzed on the best equipment, and are scrutinized by management and other senior laboratory staff before submission of results. These samples can hardly be considered representative of routine samples. However, they are used to provide evidence of capability to perform quality analyses. If the laboratory fails the PE samples, this indicates that even the most exacting analyses at the laboratory could be of poor quality. The failure of a PE sample provides evidence of the lack of capability to perform quality analyses.
- The laboratory could successfully perform in an accreditation program that includes on-site inspections and PE samples. This indicates the capability of the laboratory to perform analyses required under that accreditation or certification. However, it implies very little about whether the laboratory will analyze other samples in the same way. The accreditation might only apply to analyses of a special customer. Other samples might be processed under substantially different conditions, with different equipment, personnel and procedures.

The manager that is attempting to use current accreditation systems might now have come to the conclusion that certain accreditations and certifications could provide the information needed to determine if a laboratory is capable of performing a quality analysis on particular types of samples. Also learned is that laboratories must be required to perform the analyses using the requirements of the accreditation and certification programs to assure that the requirements are met when samples are analyzed. The manager also should have determined that inspection of the data that support the reported numbers is necessary. The manager now should understand that analytical data purchases must be inspected, as are all other critical procurements, to determine if the required quality of data is obtained.

Most managers and persons acquiring data do not ask all the questions needed to determine if the countless accreditation, certifications, licenses, and registrations are meaningful for their needs. They guess, hope, and accept data on sometimes little more than blind faith. This occurs partially because they might not understand what questions are relevant and sufficient to ask about the quality of data. Further, they might not recognize "reasonable" answers to the questions that should be asked. They might purchase data from the lowest bidder in order to justify the choice of a laboratory on a fiscal basis. Choosing a laboratory that produces quality data is difficult unless the persons procuring the data have the capability to assess and monitor laboratory quality. Therefore, only managers that have their own accreditation program can confidently purchase data at the present time. The conclusion that accreditation and certifications do not necessarily assist decision makers determine which laboratories perform quality work should be evident.

Perhaps individual decision makers are not assisted in their acquisition of quality laboratory data by current accreditation systems. However, the general public believes that accreditation, certifications, and other programs provide assurance that "bad" laboratories are not allowed to do business. Unfortunately, most of the general public probably does not worry about the issue. They rely on decision makers to exercise good judgement in acquiring data. This is not an unreasonable expectation, since the average person does not have the training or experience necessary to understand the quality requirements of data and how the multitude of accreditations assist data users in appropriate acquisition of data.

It now should be clear that the managers that create their own accreditation and certification programs have faith that they work, or are at least aware of the strengths and limitations of their own programs. It should also be evident that these separate programs are not usually considered credible by others. The current multitude of systems is not creating the means for either the general public or individual decision makers to determine if a testing laboratory is producing quality results.

In order to understand how the stated goals of laboratory accreditation may be achieved, it is necessary to study the goals of laboratory accreditation and certification by briefly contrasting the process inspection and product inspection attributes of the current systems. Detailed descriptions of the process inspection systems and product inspection systems are included in Chapter Four.

PROCESS INSPECTION VERSUS PRODUCT INSPECTION

The brief discussions presented previously about laboratory accreditation and certification programs indicate that these programs are designed to provide information about two different types of oversight:

1. Information about the process that is employed in the laboratory to produce data.
2. Information about the product of the process — the data itself.

These two types of information are both required for a good laboratory assessment program. One cannot infer anything about the actual product that is produced in the laboratory process until the data are inspected. The success of any assembly line process cannot be judged until the assembly line makes a product. This product must then be assessed according to specific requirements.

Let us consider an analogy that shows the difference and the importance of both types of oversight. Consider that you are responsible for purchase of an automobile. You can either order a car based on your inspection of automobiles from a dealer's inventory, or you can order the auto directly from the factory, based on your inspection of the factory. If you chose to purchase the auto based on inspection of cars from a dealer's inventory, you can inspect one automobile of the type that you are considering for purchase. You cannot, however, ask the dealer if he had to fix the car in any way after it was delivered to him. You also cannot inspect the factories and you cannot seek additional information about the factories. If you decide to order a car from a factory, you cannot see or inspect a finished auto, but can inspect the production facilities at the auto maker's plant. Assume that neither of the autos will have any type of manufacturer's warranty.

You can easily assess the automobiles at local auto dealers which meet your requirements, but does this mean that the automobile you order and receive will be the same quality product? What if your auto is assembled the day after the annual factory picnic — which is well known as a brawl? What if the foreman on the assembly line recently retired and the replacement begins on the exact day that your car was produced? Could either of these scenarios affect the way that your car is assembled and tested? Are any of the factories less prone to assembly of cars with defects?

Clearly, you would have a better idea of your chances of ordering a quality auto and then actually receiving a well assembled model if you knew more about the factory. Are all managers and workers well trained? Is equipment calibrated and in good working order? Are the automobiles inspected intermittently during or after assembly? Are workers given bonuses based on how many autos they produce or are they rewarded based on how many defect-free autos they produce? Do the assembly line workers follow standard routine procedures, or do they assemble each auto slightly differently each time? Are tests performed on any model to determine if the auto will perform well for one hundred miles, one hundred thousand miles, or no miles?

How many of us would order an auto if we had no warranty from the company? If we had no idea about the factory's production process, but did have one model from which to judge the quality of the factory's production process, would we order an auto? Would we assume that if the factory built one good auto, it would assemble our model correctly? Certainly, each one of us would wonder if the auto in the dealer's showroom could be considered the normal product of the factory. What exactly does "dealer preparation" mean? What are the odds of the factory assembling a "peach" versus a "lemon"?

Let us consider the converse problem. Suppose the decision was made to order the auto directly from the factory. If the purchaser of the automobile could inspect an auto-makers assembly plant, but could not inspect even one automobile, what would the purchaser know about the final product? The purchaser enters a factory and notes a well functioning group of individuals, well maintained facilities and equipment, written procedures posted on all the walls, and all functional ingredients of successful auto assembly. Does this indicate to the purchaser that a quality automobile will roll off the line? He looks for the proof — that is, an assembled automobile — but cannot see if one has been made. Would this consumer feel better about inspection of the process or inspection of the product when he buys an auto that has no warranty.

How would the individual decide which auto to buy if he had seen autos produced from one factory and had not seen any autos that were actually produced by another factory?

This consumer would probably have few qualms about purchasing one auto that could be inspected for defects at the dealer's showroom and in test drives. However, the purchaser would probably be reluctant to order an auto based on inspection of only one product from the factory. The purchaser would probably not be reluctant to order an auto if the factory that produced this type of car had been inspected firsthand and was a well run operation. Conversely, the consumer would probably have great difficulty with purchasing an automobile that he had not been able to inspect even if the factory had been inspected and seemed to have the capability to produce a quality product. Why? Even if the factory appeared to be capable of producing an automobile, the consumer would undoubtedly question the quality of the product if one were not available for inspection. In other words, what proof would a consumer have that the process could actually produce a quality product if none could be inspected?

This analogy serves to illustrate the importance of both process and product inspection. Process inspection serves to reassure a consumer that the product is capable of being produced. However, the inspection of actual product is the assurance that the required specification of product is indeed produced. The consumer judges quality in the final product. However, the consumer also recognizes that the product could not have been produced correctly unless the process worked at least for the time that the product was produced. The total process must work for the product to be constructed successfully.

EVOLUTION AND DESCRIPTION OF THE CURRENT PROGRAMS

The current accreditation, registration, licensing and certification systems have grown out of the need to determine actual capabilities and performance of testing laboratories. The growth of new and complex technologies have made it virtually impossible for customers that purchase data to easily determine which testing laboratories are capable of delivering quality data. Therefore, various types of accreditation bodies and systems have evolved as attempts to assure quality data through regulatory, contractual, and other similar approaches.

Accreditation and other similar attempts to determine capability and performance have been in existence for years for a variety of disciplines. There are, in fact, so many programs, that even the naming of the programs currently causes confusion. The naming of many of the current programs is often misleading because what one body calls accreditation could be termed certification by another program. Several chapters in this book describe accreditation, certification, registration, and other systems in detail and clarify the basic requirements of each system. The confusion that currently exists about the merits of each program lies, in part, in the confusing nomenclature and the public's assumptions of what the names of the systems connote about the testing laboratories. This subject is covered in detail in Chapter Two.

There is a strong basis for accreditation and certification in almost all disciplines and in the manufacture of many products. Some accreditations involve approval by an evaluating body such as the accreditation of law schools by the American Association of Law Schools and the American Bar Association. The Environmental Protection Agency certifies laboratories performing drinking water analyses. States license professional engineers. Third-parties list and certify products. All such systems and parties seek to establish the competence of persons and establishments, or performance of products and services.

Do any of the accreditation and certification types of systems assist consumers in determining what establishment will deliver quality products and services? Not one of the systems can assure consumers that they will actually get quality products or services from the establishment, but some of the better systems do reflect the establishment's capability to deliver quality goods and services. What is needed is a system of continuous inspection of the products and services to assure consumers that each time a product or service is delivered it will be high quality.

Continuous, ongoing monitoring of goods and services is needed to assure consumers that establishments deliver quality products. This statement allows the reader to easily understand why some accreditation systems are more successful than others. Consider a licensed professional engineer that designs buildings that do not pass city building codes. Consider an accredited law school that graduates students that consistently fail to successfully perform on the state bar association examinations. Consider the certified product that fails after installation in computers, and causes ten percent of all computers assembled in a plant to be returned. In these cases, it is easy for the consumer to recognize failure of the persons or establishments that produced unacceptable results. It is also easy to determine that the accreditation or certification systems that did not discover the incompetence or problems also failed. Continuous oversight of these systems is easily provided by consumers that use the products and services that have been produced by certified or accredited entities.

Consider the certification program for drinking water laboratories. Can consumers ever determine if the laboratories in the system are producing inferior services? Private individuals rarely ever have their own water tested, but even if the private individuals were given imprecise and inaccurate assessments of their well water quality, would these persons be able to judge the quality of the results? The answer, is in most cases, no. Private individuals might obtain invalid numbers that would be formatted and delivered exactly like valid numbers. How then, are incompetent drinking water laboratories or a poor drinking water certification system ever to be discerned? The reader can easily ascertain that the more successful accreditation and certification programs are those that can be easily monitored by many people. Analytical laboratory certification and accreditation systems cannot be readily assessed by the general public. Therefore, they do not assist data purchasers in this example or in related instances of purchasing data.

Analytical laboratories perform many types of tests. These tests range in difficulty and in the types of equipment and analytical training that is required for successful analysis of samples. Clinical and toxicology testing laboratories have been accredited or certified by many states for a number of years.[5] This fact seems to suggest that these accreditation programs are viable, even though patients usually do not know what test the laboratories are performing on their samples. Does this mean that the customers are able to determine whether the data produced by the laboratories is shoddy or quality work? No, this means that competent physicians, as the patients' agents are capable of determining if laboratories are supplying reasonable work. Human samples can only have a certain range of analytes in the samples. Extreme outliers in the data would be noted as impossible to find in human samples. In addition, if the numbers indicate serious problems, and the samples provided the only indication that problems existed, the physicians would most certainly question the test data as a possible spurious laboratory results and order retests. The physicians use the results, in conjunction with other patient information. The physicians would not rely on the laboratory data alone as the basis of a decision. It is easy to see that physicians will chose an accredited laboratories based on assessments of their capabilities by outside authorities. The physicians, however, provide continuous assessments of actual laboratory performance by noting how many questionable results the laboratories deliver. Needless to say, if physicians believe each laboratory test without ever questioning the numbers, the patients are not receiving the best care. The patients rely on educated agents to assess the laboratories and therefore the accreditation systems.

Consider the previous example used concerning certification of drinking water. If municipal water authorities receive data from laboratories that indicate that the water is seriously contaminated with a material that had never before been discovered in the water, the laboratories would most certainly be questioned about the result. The fact that the laboratories are certified would not mean that a mistake could not be made. Public water authorities would expect to see the data reflect what is historically reasonable. Any spurious result would be cause for alarm and would be promptly investigated. The sample would probably be reanalyzed to determine if the first sample result was correct. For this reason, a constant oversight system for drinking water laboratories is in place. However, this oversight system might only detect those data that are cause for public concern. A more serious concern is the problem that could exists if some of the laboratories are incapable of finding a serious contaminant in water. In fact, this contaminant could be in the water supply and never be discovered only because the laboratories in which the samples are analyzed are not capable of successfully analyzing for it. In this case, the constant oversight system of the public water authorities would fail to find the most serious type of error.

The preceding paragraphs described many types of accreditation and certification systems to show that a myriad of systems exist. These systems range from effective to ineffective. Let us assume that the requirements of each system are effective at prediction of capability to perform. What makes some systems effective and other systems ineffective is some other factor. In the several examples presented, it has been shown that continuous monitoring and verification of ability may be the key factor. If this is indeed the case, then all successful systems must have continuous monitoring, whether it is performed by the accrediting body or by informed consumers or their agents. The following chapters of the book will provide support for this statement.

THE ROLE OF THE GOVERNMENT IN ACCREDITATION AND EVALUATION

The goal of all accreditation and evaluation systems is to verify that an establishment is capable of providing quality goods or services. In some cases, this verification is considered so important that federal and state officials have enacted legislation that requires accreditation and certification of specific services and products. Clinical and toxicology laboratories have been regulated by such legislation for years.[6] The Safe Drinking Water Act granted primacy for laboratory accreditation to the 50 states in 1978. All these systems were established because the government believed that they were critical to protect the public health and the environment.

The government's role in accreditation and evaluation systems is to implement a system that determines which testing laboratories are capable of producing quality data for government decisions. Further, as in any other case where the public needs the government's assistance in protecting the public interest, it is the government's responsibility to assist the public in determining which laboratories are capable of performing acceptably on samples that affect the public. It is also the government's role to protect the public from establishments that are not capable.

How does the government exercise its responsibility to protect the public health and environment in the current systems? In some cases, state legislation is enacted to protect the citizens of each state. These regulations sometimes differ substantially from state to state. The differences should be critically inspected at the national level to determine if the states are all adequately protecting the public health and environment. In many cases, state requirements are more stringent than national requirements. In these cases, the problem is not that the requirements do not adequately protect the public and the environment. The problem is that differing requirements could cause confusion in laboratory operations. The confusion could impact laboratories' ability to produce the required quality data. These differing requirements may sometimes be of little consequence to local state laboratories that only perform analyses on samples from the state in which they are located. The conflicting requirements for analyses and requirements for certification in many states becomes burdensome for laboratories that analyze samples that originate from many states. Differing state requirements have a similar impact on national clinical laboratories since testing at these laboratories could be performed on samples that originate from many states.[7] This same problem has been partially resolved in the clinical laboratory industry with enactment of the Clinical Laboratory Improvement Act Amendments of 1988.

Are the redundant and conflicting requirements and membership in several different accreditation and inspection programs only a laboratory problem? Is this unfortunate circumstance of the many systems that states have mandated to protect the public health and the environment only a business problem? Does the federal government have an obligation to make it easy for laboratories to do business? Does the government have an obligation to increase the profitability for independent laboratories? One might argue that it is not the government's responsibility to decrease needless redundancy and cost in laboratory work.

However, the government has a responsibility to acquire quality data in a cost-effective manner. The government is one of the largest purchasers of laboratory data. The cost for redundant and ineffective quality assurance (QA) systems is passed on to the general public because the taxpayer pays for the government's laboratory data. The public not only bears the cost of development and implementation of redundant government assessment and certification systems, but also pays more for the data that is produced in the laboratories. The government does have an obligation to spend the taxpayer's funds wisely in their quest to protect the public health and environment. It is easy to see that in assisting the private sector to most effectively produce goods and services, the data that the government purchases will become less costly. Therefore, the tax burden from purchase of data will be lowered. Judicious use of public funds is the government's responsibility. Therefore, it is the government's responsibility to assist in drafting and enacting national regulations that allow for a cost effective means to assess the quality of data and to prohibit redundancy in such programs.

THE ROLE OF PROFESSIONAL AND TRADE ORGANIZATIONS

Professional and trade organizations have long sought to assist their particular trades and professions in differentiating the good from the bad. Responsible professionals and establishments want to be distinguished from counterparts in their trades and professions that deliver lower quality products and services. It is in the best interest of these individuals and reputable establishments to request the assistance of their trade or professional organizations to assess the trades and professions and accredit, certify, license, or register those that meet the required criteria.[8]

Trade and professional associations have enlightened self-interest at heart when they sponsor evaluations of their members. It is in any individual industry's or profession's best interest to improve their public perception through self-inspection. If they improve their profession or trade as a whole, they have assisted each member.

Do trade organizations and professional organizations have a responsibility to their customers to implement accreditation or evaluation procedures? Do trade organizations have a responsibility to their customers to assist the government in the design and implementation of the government's accreditation and inspection programs that will regulate their trade or organization? The answer to both of these questions is a resounding yes.

It is appropriate that the individuals that collectively make up the trade organizations increase customer awareness of the quality of their trade or profession. It is the trade organization's responsibility to assist customers in recognizing quality goods and services and protect the trade or profession's integrity by avoiding the delivery of inferior products. The government might lack the in-depth expertise in a particular trade to design the most effective accreditation or inspection system if the industry does not assist in the effort. The trade or profession should endeavor to assist the government's design and implementation of a system that both protects the consumer from inferior goods, but also is not unreasonably burdensome to the industry. Ideally, the implemented system should both assist the consumer in their selection process and improve the trade or profession by culling out the poor performers.

Unfortunately, in many instances, the government is reluctant to accept the assistance that industry offers. In such cases, government officials are attempting to protect their systems from the trade and professional associations. The assumption is that the organizations will attempt to dilute the full force of requirements by making the requirements less demanding on their members. This might happen upon occasion, but the effect of making requirements less stringent would actually not be in the trade or profession's best interest. The trade organization and the government are both interested in identifying the individuals and establishments that are not capable of performing. The trade associations are interested in including the requirements that can be implemented without undue burden on their members. They are interested in establishing the fewest, most effective requirements. If, in fact, the trade association could assist the government in making its systems more effective and streamlined, this assistance should be welcomed. In those instances where government officials have reason to believe that associations are attempting to dilute the effectiveness of the system, their advice will be rejected. Assistance from trade associations should never be discarded before it is carefully evaluated for possible merit. These organizations represent the trade that will be monitored, and should therefore have considerable insight into the issues that surround the proposed accreditation and certification systems.

THE ROLE OF THE LABORATORY INDUSTRY

Accreditation can be of great benefit to the laboratory industry. A good accreditation program assures clients that individual laboratories are capable of producing quality data. If the public has confidence in the accreditation system, then the public will have greater confidence in the laboratory industry as a whole.

Because the laboratory industry stands to gain so much from a credible accreditation program, it would seem that the laboratory industry would eagerly participate in the development and implementation of such a system. However, the laboratory industry, as a whole has been reluctant to support accreditation. The cause for this reluctance is simple. A myriad of often conflicting and usually redundant accreditation systems is already in place. The current situation almost guarantees each laboratory that it can be accredited or certified by at least one program. Also, a single reasonable and effective accreditation program is a mirage to the laboratory industry. If a single, effective, technically sound accreditation program for testing laboratories could become a reality, it would be supported by many laboratories.[9] However, the laboratory industry does not believe that this situation will be possible soon, based on slow progress in this effort in the past.

Another reason that some laboratories are reluctant to support a national accreditation program is because so many of the existing certification programs do not have credibility. They do not wish to support a program that will provide assurance to data purchasers that laboratories are capable and further, that quality data is provided, when the certification or accreditation body's program is not capable of providing that assurance. They fear that a national laboratory accreditation system would fail to be rigorous enough to be better than the existing systems that do not provide assurance of quality laboratory data.

Even if all laboratories are not in total support of a national program, they all agree that they are burdened by a chaotic approach to accreditation that ineffectively uses valuable QA resources. The time and effort that could be expended on assisting regulators make order out of the chaos are currently wasted on redundant and costly accreditation programs. Certainly, one cannot blame the conscientious managers that have designed and implemented systems to provide assurance that the data that they procure is of the required quality for this situation. They are, in fact, performing their jobs as well as they can. However, they might not be knowledgeable about the requirements of other systems, the redundancy that exists with other programs, or the costs of the systems to the laboratory industry which are ultimately passed on to consumers.

Laboratories can assist in minimizing the profusion of programs and minimize the waste of resources that the multiple systems cause. Who better can point out the waste in the profusion of programs than the laboratory industry? It is the laboratory industry's responsibility to produce quality data at a reasonable price for consumers. It is therefore the laboratory industry's responsibility to improve the accreditation situation as best it can by alerting government agencies and other clients of problems and possible solutions.

Laboratories have a responsibility to assist in the development and implementation of accreditation and evaluation programs that will ultimately improve the laboratory industry and the quality of the data that this industry produces. The laboratories must assist regulators at the state and federal level in order that the expertise and knowledge that resides in the laboratories can improve the current situation. The laboratory industry should support their industry's trade organizations to promote a credible national laboratory accreditation program. It is in the best interest of the individual laboratories, the industry as a whole, and the purchasers of laboratory data for a credible national laboratory accreditation system to be implemented.

THE ROLE OF PRIVATE CERTIFICATION COMPANIES

Private certification companies have been developed to meet a perceived need by the public and industry alike. In many cases, the public believes that a trade association or professional organization cannot effectively inspect one of their members and ever find a member who is not complying with all requirements. Industry sometimes has welcomed the use of a competent third party to perform inspections that are required by regulatory agencies. The laboratories wish to comply and benefit by providing customers with evidence that an outside third party has found them in compliance with the requirements.

What is the best role of the third-party companies? Perhaps the most

critical role is not merely to inspect to industry or government requirements, but the third parties might also serve another important role.
This role could be the development of consensus standards that delineate the criteria and requirements for accreditation and certification.
Some third-party companies have performed this service in developing
product standards.[10] These companies might be able to assist the various
state and federal regulators, the industry representatives, and the public
sort out the technical, legal, and political requirements of the current
systems. The result can be a uniform, national program that will ultimately reduce cost and increase the quality of data produced in the
laboratory industry.

What responsibility do these third-party companies have? These
companies must remain unbiased and maintain a high level of integrity
to all parties that seek the establishment of consensus standards. They
have the responsibility to have the financial stability and technical infrastructure to maintain, update and clarify the standard to best suit all
parties at the request of any of the affected parties. They have the
responsibility to maintain a no-conflict-of-interest requirement with
respect to all the parties involved with the standard. Further, the third
party that develops the standards has the responsibility to assure that any
requirements for implementation of the standards can be accomplished
by any credible third party.

STATEMENT OF THE PROBLEM: CURRENT PROGRAMS DO NOT EFFECTIVELY MEET THE GOALS

The goal of all accreditation, certification, licensing, or registration
programs is to assure data users that a laboratory is producing quality
data. Current programs do not effectively meet this goal. Laboratories
and users of data alike agree that a credible accreditation program is
needed. Even though several accreditation and certification programs
currently exist, the problem of how to determine if the work of a
laboratory is credible still exists. In fact, the Environmental Protection
Agency's report to Congress[11] on the Availability, Adequacy, and
Comparability of Testing Procedures for the Analysis of Pollutants
Established Under Section 304(h) of the Federal Water Pollution
Control Act listed as a major finding:

> No Agency guidance exists for certification of laboratories performing
> analyses under provisions of the Federal Water Pollution Control Act

(FWPCA). Independent laboratory certifications for waste water analyses are performed by 18 states. Comments received from representatives of the regulated community on the draft of this report strongly supported implementation of a nationwide, federally-administered laboratory certification program.

Further, based on that finding, the recommendation was that the Agency should take the following action:

Consider the establishment of an Environmental Monitoring Management Group to coordinate the development of uniform test procedures and quality assurance/quality control guidelines within the Agency, to provide a long-range program for the development of analytical methods and to explore the feasibility of a uniform, national certification program for laboratories performing test procedures.

This group has been formed and is charged with the above listed actions.[12]

Even the EPA's own federally funded regional laboratories are voicing complaints about the number of on-site visits and performance evaluation samples that they must participate in to work for various EPA program offices. They are suggesting consolidation of the various federal EPA laboratory evaluation programs because of the burden these federal programs have on EPA regional laboratory resources.

Is accreditation or certification of laboratories for the purpose of establishing credibility in laboratory data needed? The answer is an emphatic yes. The adequacy and accuracy of laboratory tests are critical for correct decision making. It is imperative that these data be the required quality for decision making. It is imperative that the laboratory community be assessed in such a manner that the competence of testing laboratories to produce quality data is determined.

THE SOLUTION — AN INTRODUCTION

In the following chapters, the basic requirements of accreditation, certifications, and other evaluation programs will be examined in order to provide the basis for the System for Success. The System for Success is a merging of several proven ways to assess and assure quality. It can provide the basis for a national laboratory accreditation and certification program and help provide the answers to the current accreditation dilemma.

It is important to stress that all of the key elements of the System for Success must be implemented for accreditation and certification to have value. If the System does not include capability inspection as well as performance monitoring, the System is worse than useless. A poorly designed accreditation or certification system is worse than no system at all because it leads data purchasers to believe that certified and accredited laboratories produce quality data. In fact, most systems that currently exist reduce data purchaser's chances for obtaining quality data. If there were no accreditations or certifications, data purchasers would need to determine if data produced from laboratories met their expectations, rather than relying on an accreditation to assure them of quality. The evaluation of data is vital, and must be accomplished by some entity for laboratory accreditation and certification to have value to data purchasers.

It is also important to stress that laboratory accreditation and certification would not be necessary if all data purchasers worked with laboratories in purchaser-supplier relationships. If laboratories and data purchasers work cooperatively, the quality of data can constantly be verified by both the laboratories and the data users. However, since this is not the case in many instances, data purchasers seek tools to assist in their evaluations of laboratories. One such tool can be a well designed accreditation and certification program that evaluates both capability and performance. The System for Success is designed to assist data purchasers determine if a laboratory produces quality data. It is also designed to assist laboratories in providing evidence of quality to data purchasers. Only when both laboratories and data purchasers understand the basic requirements for generating quality data and work together will the best data for the intended purpose be generated. The System is designed to assist both laboratories and data users work together to this common goal.

REFERENCES

1. "Availability, Adequacy, and Comparability of Testing Procedures for the Analysis of Pollutants Established Under Section 304(h) of the Federal Water Pollution Control Act, Report to the Committee on Public Works and Transportation of the House of Representatives, EPA/600/9-87/030, p. 1.1 (1988).

1a. Schneider, Keith, "Faking It — The Case Against Industrial Biotest Laboratories, *Amicus Journal*, pp.14-21 (Spring 1983).

2. Code of Federal Regulations, Title 21, Part 58, 1978.

3. "Envirite Fined for Lab Violations," *Superfund* 4(6):8 (1990).

4. "DOE Internal Audit Provides Details on Compliance Failures," *Hazardous Materials Control Research Institute - Focus*, p. 4 (September 1989).

5. "Laboratory Accreditation — Requirements Vary Throughout the Federal Government," Report to the Chairman, Committee on Science, Space, and Technology, House of Representatives, U.S. General Accounting Office, GAO/RCED-89-102 (1989).

6. "Clinical Laboratory Improvement Amendments," U.S. Public Health Service Act, Public Law 100–578, Section 353 (1988).

7. McCarthy, Carolyn Kim., Meeting Presentation at the International Association of Environmental Testing Laboratories Accreditation Subcommittee Meeting, Wilmington, DE (February 6, 1989).

8. Felsen, Harvey G., "What's Wrong With Our Industry?" *Environmental Testing Advocate* II(1): 1,3 (1990).

9. Proceedings of the USATHAMA Quality Assurance in Environmental Measurements Meeting, Prepared by Science and Technology Corporation, Hampton, VA (May 1989).

10. Chumas, Sophie J., Ed., *Directory of United States Standardization Activities*, National Bureau of Standards Special Publication 417, U.S. Government Printing Office (November 1975).

11. "Availability, Adequacy, and Comparability of Testing Procedures for the Analysis of Pollutants Established Under Section 304(h) of the Federal Water Pollution Control Act, Report to the Committee on Public Works and Transportation of the House of Representatives, EPA/600/9-87/030, pp. ix–x (1988).

12. "Instrumentation '90" *Chemical and Engineering News* 68(23): 26–32 (1990).

Chapter 2

CERTIFICATION, ACCREDITATION, REGISTRATION, AND LISTING

DEFINITION OF TERMS

Many terms are used to describe the various programs that assess laboratories. The terms are used almost interchangeably in describing evaluation programs. This use of many different terms has added to the current confusion surrounding accreditation and certification. What are the definitions of these and other related terms that are commonly used to describe accreditation and certification-like systems? Can they correctly be used interchangeably, or are they different entities? In an effort to better describe what these terms mean, the following definitions from Webster's 9th Collegiate Edition[1] are quoted:

> Accredit — *vt* 1: to consider or recognize as outstanding 2: to give official authorization to or approval of: a: to provide with credentials; *esp* : to send (an envoy) with letters of authorization b: to recognize or vouch for as conforming with a standard c: to recognize (an educational institution) as maintaining standards that qualify the graduates for admission to higher or more specialized institutions or for professional practice 3: ATTRIBUTE, CREDIT *syn* see APPROVE.

> Certify — *vt* 1: to attest authoritatively: as a: CONFIRM b: to present in formal communication c: to attest as being true or as represented or as meeting a standard d: to attest officially to the insanity of 2: to inform with certainty; ASSURE 3: to guarantee (a personal check) as to signature and amount by so indicating on the face 4: CERTIFICATE, LICENSE.

> *syn* CERTIFY, ATTEST, WITNESS, VOUCH mean to testify to the truth or genuineness of something CERTIFY usu. applies to a written statement

esp. one carrying a signature or seal; ATTEST applies to oral or written testimony usu. from experts or witnesses; WITNESS applies to the subscribing of one's own name to a document as evidence of its genuineness; VOUCH applies to one who testifies as a competent authority or a reliable person and who will defend his affirmation *syn* see in addition APPROVE.

Inspect — *vt* 1: to view closely in critical appraisal : look over 2: to examine officially (inspect the barracks every Friday) *syn* see SCRUTINIZE .

Registered — *adj* 1 a: having the owner's name entered in a register (registered security) b: recorded as the owner of a security 2: recorded on the basis of pedigree or breed characteristics in the studbook of a breed association 3: qualified formally or officially.

Qualified — *adj* 1 a: fitted for a given purpose : COMPETENT b: having complied with the specific requirements or precedent conditions for an office or employment : ELIGIBLE 2: limited or modified in some way (qualified approval) *syn* see ABLE.

Approved — 1 *obs* : PROVE, ATTEST 2: to have or express a favorable opinion of (couldn't approve his conduct) 3 a: to accept as satisfactory (hopes she will approve the date of the meeting) b: to give formal or official sanction to : RATIFY (Congress *approved* the proposed budget).

syn APPROVE, ENDORSE, SANCTION, ACCREDIT, CERTIFY mean to have or express a favorable opinion of. APPROVE often implies no more than this but may suggest considerable esteem or admiration; ENDORSE suggests an explicit statement of support; SANCTION implies both approval and authorization; ACCREDIT and CERTIFY usu. imply official endorsement attesting to conformity to set standards.

License — *vt* 1: to issue a license to 2: to permit or authorize esp by formal license.

To clarify the above definition, one of the first definitions of the noun license is given:
1 License — *n* 2 a: a permission granted by competent authority to engage in a business or occupation or in an activity otherwise unlawful b: a document, plate, of tag evidencing a license granted.

List — *vt* 1 a: to make a list of : ENUMERATE b: to include on a list : REGISTER 2: to put (oneself) down 3 *archaic* : RECRUIT.

The dictionary definitions of the terms show that the meanings of these terms are significantly different from each other. Using the terms interchangeably in describing laboratory assessment programs is incorrect. It can mislead the public and purchasers of data into believing that the terms correctly describe the assessment programs and that the dictionary definitions of the terms apply to laboratory assessment programs. This state of confusion has become so widespread that the terms are used interchangeably in documents.[2] Even though international agreement has been reached that accreditation should be applied to organizations and certification should be used for individuals and products,[3] the terms still are often used synonymously.

Certification can best be described from the dictionary definition as attesting to the truth. Accreditation can best be described as attesting to reputation, as in meeting official requirements. Licensed means to have permission from authority to engage in a business or occupation. Therefore, the term, license, does not convey any special meaning about the truth or reputation of the business or occupation. It merely states that the laboratory has been licensed to do business. This could mean that a fee has been paid, and in such a case would not connote assessment of the business or occupation.

The terms listed and registered also do not convey any meaning about truth or reputation. Listed does not imply any information about the credibility or competence of the entity listed. The term means only that the entity is included on a list. Some programs such as the Underwriters' Laboratories (UL) and the National Sanitation Foundation (NSF) use the word "list" to connote assessment and approval of the quality of products.[4,5] This use of the word is not the usual; therefore, consumers must be educated about this use of the term before "listed" products are recognized as different from other products.

USE OF THE TERMS ACCREDITATION AND CERTIFICATION

The dictionary definitions of many of the defined terms differ substantially from many of the descriptions of these terms when used by professional organizations, trade associations, third-parties, quality assurance professionals and the government. Some examples of how these terms are used and their sources are provided in this chapter. The programs that use these terms are described in detail in the following chapters that deal specifically with the different types of systems.

- Certification as described by the proposed European Certification System[6] is the assurance of satisfactory product performance. The certification culminates in the "CE" mark being placed on products. The process leading to placement of the certification mark on products includes:

1. Recognition of standards
2. Compliance testing and inspection by an approved laboratory to determine product compliance with the standard requirements
3. Plant inspection
4. Coordination of the results of the laboratory product testing and inspection and the plant inspections.

This last step of coordination is the third-party certification step that enables manufacturers to receive "CE" labels for their products. The "CE" mark then conveys that the product and plant have met the certification requirements.

- Accreditation is a means frequently used to identify competent testing laboratories. It implies that "a laboratory has been evaluated by a recognized authority and found capable of conducting specific activities." This description is a direct quote taken from a paper included in the American Society for Testing and Materials (ASTM) Special Technical Publication 814.[7] The author of the paper presented activities of the ASTM Committee E-36 in standards development for laboratory evaluation and accreditation.

- Accreditation is described by the National Voluntary Laboratory Accreditation Program (NVLAP)[8] as having the objective of providing national and international recognition of competent laboratories. NVLAP officials point out that the approach taken by accreditation systems is critical in determining if the laboratory's clients can have an adequate basis for confidence in the laboratory's competence in the test methods that are to be applied. They state that NVLAP accreditation is granted for specific test methods within specific areas of testing. This "accreditation by test method" provides confidence in the test methods. The program was set up by test method in the belief that other "accreditation by general field of testing" does not provide an adequate basis for confidence in the laboratory.

- Accreditation is described by three quality assurance professionals in the publication of the American Society of Quality Control.[9] They state that "Laboratory clients are naturally interested in the quality and reliability of purchased laboratory data. At the same time, laboratory management is interested in demonstrating its lab's competence. Laboratory accreditation systems have been established to meet both of these needs by ensuring that accredited labs produce accurate data and comprehensive reports." They further state that "It is assumed that a laboratory complying with the accreditation system's criteria will be doing the testing properly and obtaining accurate data."

- Accreditation as distinguished from certification has been described by the Chemical Manufacturers Association (CMA).[10] They describe confusion in the current use of the terms. They state that accreditation, by convention, has been applied to organizations such as laboratories and that certification should be associated with individuals. They further recognize that the word certification is currently used most often in describing laboratories. They suggest that the term accreditation is not universally defined, but one definition is "the process of verification by competent individuals that a laboratory possesses the capability for providing precise and accurate data on a day-to-day basis and that the data can be routinely relied upon to meet data quality objectives."

- Accreditation has been described by a spokesperson and quality assurance professional[11] in charge of quality assurance for a national network of laboratories as "Accreditation programs have been established to provide a measure of assurance to the users of environmental data, whether they be the regulators or those being regulated, that the laboratory producing the data is qualified to generate that data." Further suggested is that no accreditation programs currently in operation actually provide a framework for assuring overall data quality. Then suggested is that a national certification program that provides certain reportable elements to the end-user of the data would better meet the stated goals of the accreditation definition.

- Accreditation has been described by a polarized light microscopy (PLM) professional[12] as one means to "take some of the guesswork out of selecting a PLM laboratory." He adds that there are other practical considerations that are needed in the selection of a laboratory. He states that even though there are many requirements that are checked by the National Voluntary Laboratory Program (NVLAP) which is implemented by the National Institute for Standards and Technology (NIST), "the most critical component in a high-quality PLM lab, however, is the analyst."

- Accreditation is described by the director of the American Association for Laboratory Accreditation (A2LA)[13] as essentially a third-party appraisal of a laboratory's capability to perform quality work.

- Certification is sometimes defined as "official approval" by different organizations. This type of definition is used by the Environmental Monitoring Systems Laboratory - Cincinnati (EMSL-Cinn),[14] the EPA laboratory that responds to the technical needs of the EPA Office of Drinking Water:

  > "Certification is defined as the official EPA approval by a regulating authority based on exact Agency criteria, for a laboratory to perform some required function. This definition is used by the EMSL-Cinn to differentiate certification from accreditation. To further clarify the difference, the Agency states that there is only one Certification Program in EPA, and that is the one for certification of laboratories regulated under the Safe Drinking Water Act (SDWA) and its amendments. This program certifies laboratories that analyze drinking water."

- De facto certification is the term used by industry and some EPA officials to describe Superfund's Contract Laboratory Program (CLP).[15] The Agency has stated that "With the constant monitoring for lab performance and the high level of QA/QC, the CLP reputation gives Agency programs a strong "comfort factor" choosing CLP laboratories for other analytical services. The CLP has become a de facto certification program for the laboratories performing analysis of waste samples."

Third-party certification is defined in the American National Standards Institute (ANSI) Z341-1987 as, "A form of certification in which the producer's claim of conformity is validated, as part of a third-party certification program , by a technically and otherwise competent body other than one controlled by the producer or the buyer."[16]

CURRENT USE OF RELATED TERMS

Registration and Licensing of Individuals

The term registration is difficult to understand because it varies in its meaning and is sometimes linked with other terms, such as licensing. It can be used exactly like the words accreditation and certification to show that individuals have met stringent sets of criteria that provide evidence that they are qualified in a trade or profession.[17] The term, used in this sense, is used to show that individuals have been formally qualified by a process. However, some systems allow professionals that have been working in the profession many years to be "grandfathered" or registered in a program without undergoing the usual, formal qualification processes.

Some registration programs provide more oversight of the performance of their program than others by investigating complaints from the public about their registrants. The complaints are investigated in order to protect the public and to enhance the profession. Other registration programs might not have formal complaint investigation procedures.

In some cases, professional registration boards might be linked together with state licensing programs. In such cases the professionals must both meet certain professional requirements such as successful performance on an examination and also pay fees to state registration boards. The professionals are then licensed by the state. In some states, certain professionals might be required to be licensed in order to perform professional work in those states. An example of this case are professional engineers who are registered by professional boards, but must pay license fees in some states in order to conduct professional engineering business in those states.

Consumers should have an understanding of what qualification processes entail in order to have confidence in individuals that are registered or licensed. Registration of individuals, in many cases, is a procedure that is rigorous and should be capable of determining which individuals are capable of successfully performing in the particular trade or profession. However, some registration of individual programs are not as rigorous as others. As has been discussed, the requirements behind the different programs differ substantially from program to program. Because this is the case, consumers must look into each program individually to determine what the terms mean. In effect, the consumer must become informed about the basic requirements of individual registration programs to determine if they are rigorous and provide meaningful information about the capability and ability of the people that are registered in the programs.

Cost and Benefits of Registration and Licensing of Individuals

Registration costs the individual professionals that are registered. The fees for registration, licensing, examinations of individuals, investigations of complaints, and all other registration expenses are paid by the professionals that are licensed in the states. In some states, the state general fund also gains revenue from the professional members. In such cases, the state gets part of the working budget of the registry and licensing and adds these funds directly to the state budget.[18] In such cases, the states are adding a tax to the professionals that do business in those states. (The fees may be used to support programs of direct benefit to the professionals in some states.) The consumer is likely impacted by the added cost to professionals.

In many cases, individual professionals are required to pay for their own registrations and licenses and are required to be licensed to keep their positions. It is easy to see why some states would not want the several types of alternative taxation schemes and state revenue boosting evaluations, registrations, licensing fees, and accreditation schemes to be eliminated. The state general fund may be increased by many of these operations, to a greater extent than the programs cost the state. The cost of many registration programs would probably not need to be as high as they currently are if added taxation were not added to the fee structure in some states.

The public is helped by many registration programs, even if they do not directly pay for the services. The good registration programs benefit the general public because they minimize the probability of professionals who perform shoddy work by testing the individuals against criteria to determine which individuals are capable of performing good work. They also investigate complaints about their members. Many registration programs also offer educational opportunities and other professional help to improve the competence of the registered professionals.

Public Perception and Credibility of Individual Registration and Licensing

Some registration programs are much more credible than others. The terms registered and licensed sometimes converge and might be the source of some confusion. In some states, registered professional engineers or registered professional land surveyors are licensed by the state and their work is investigated by the state if a complaint is filed. In many states, licensed professionals must sign that work is performed correctly in order for the work to be recognized as adequate by the state. In such cases, the credibility of the system is judged over the long-term by the quality of the work that the professionals "sign-off" on.

Some registration programs such as those that register professional engineers are well known to the consumers that use their services. Some of these programs are trusted and others are not held in high esteem by consumers. The consumer sometimes bases trust in the programs because of a real understanding of the requirements of the registration programs. In other cases, the trust might be unfounded; however, experience or other information causes the consumer to trust registration programs. In these cases, the registration programs might be capable of discriminating between individuals that meet and those that do not meet the requirements of the registration program.

In most registration systems, the professionals have distinctly different skills, based on both education and experience differences. The registration systems might be able to indicate baseline conformance to requirements, but probably cannot differentiate between very good and marginal registrants.

One reason that some professional registrations might be considered suspect by consumers is because many registration boards rarely remove members from the registry because of complaints. These registration boards are viewed as professional societies. This type of registration entity is lenient in expelling members. One the other hand, some registration boards are stringent; many individuals are investigated and are removed from the registry if they are found to be negligent in their professions. Public perception of registration is therefore mixed based on personal and substantial knowledge of the registration program and its requirements.

Registration of individuals is different from all other systems to determine quality. These registration programs do not check for compliance of a product to requirements, but measure the qualifications of individuals against a set of requirements. These requirements are designed to enable one to predict if individuals are capable of performing quality professional work. The assessment procedures many times include written examinations. Most registration programs do not inspect an actual professional work product unless a complaint is charged against a registered professional. In such cases, the professional registration board could investigate and if the work product is determined to be defective, the registration body could take appropriate remedial action. In such cases, the registration body does inspect product, but only as a corrective action. The intent of such registration programs is to establish capability, much in the same manner as accreditation programs establish capability for organizations to perform good work.

Registration of Processes

The term registered is also used to describe the process in which quality assurance programs are audited and registered by third party assessment and registration agencies. The meaning of the term registration, if used this way and if the registration programs are rigorous, is synonymous with the word certified as it is used to describe the certification of vendors and vendor quality systems.

Quality assurance registration programs have recently become better understood since just-in-time management procedures have promoted the prevalence and understanding of the reasons to conduct vendor quality assurance registration programs. As is the case with all QA programs, registration programs provide results that are commensurate with the requirements that govern the registration processes.

Description of Current Process Registration Programs

The U. S. Food and Drug Administration (FDA) uses the term register in their regulation that applies to the manufacturers and importers of finished medical devices intended for human use. They require the manufacturers and importers of medical devices to register with the FDA and comply with a Good Manufacturing Practice (GMP) regulation.[19,20] In order to determine if all companies that have registered with FDA as medical device companies are in compliance, the FDA periodically inspects for compliance with the GMP regulation. The FDA GMP regulation was "developed in accordance with basic principles of quality assurance". It states that "these principles have as their goal the production of articles that are fit for their intended uses, and can be stated as follows: (1) quality, safety, and effectiveness must be designed and built into the product; (2) quality cannot be inspected or tested into the product; and (3) each step of the manufacturing process must be controlled to maximize the probability that the finished product meets all quality and design specifications." Registration with FDA as a medical device manufacturer therefore denotes compliance with a rigorous quality assurance program from the original design of the device through the manufacturing process to the final inspection for conformance to requirements.

The FDA program also applies to contract manufacturers of subassemblies and to manufacturers of accessories or components that are packaged and labeled for a health purpose for the end user. These manufacturers do not have to register with FDA, but primary registered manufacturers that use such suppliers must assure that their suppliers comply with the GMP regulation. This regulation, therefore, requires primary manufacturers to purchase supplies from manufacturers that make subassemblies in accordance with good basic quality assurance principles. The importance of this requirement for quality requirements of a primary supplier to also extend down to secondary suppliers is a key element of any good quality assurance program. The importance of this element is explained in Chapter 6.

Cost and Benefits of Registration and Licensing of Processes

In the future, the concept of registering quality systems will become more important. The European Economic Community (EEC) will adopt a registration process in l992 for imports and exports in order to establish a consistent level of quality. The ISO 9000 (ANSI/ASQC Q 90 series)[21] standards are expected to be the basis for the new registration requirements. The Europeans have determined that quality system registration is needed to assure the international marketplace that a consistent level of quality is achieved. One of the elements in their system to achieve consistent quality is the registration of quality systems.

Single inspection and registration of processes and quality assurance systems so that the processes and systems do not need to be inspected by each individual that needs an assessment is a cost effect means to provide inspection and quality assurance oversight of processes, quality systems, laboratories, and any other system that can be treated in this manner. The EEC system was developed to establish and provide consistent quality without redundant and costly inspections. Registrations are a cost effective way to establish quality that benefits both the inspected and registered entity and all entities that use the registration program to establish quality of systems.

Public Perception and Credibility of Process Registration

Product and process registration most certainly are not recognized to be similar to product certification or listing that are described in Chapter 4, even though some systems are similar. Registration of quality programs or manufacturers can entail both accreditation of process to determine capability and also inspection of product to determine product conformance. The term registration can also be used to show that the company is listed and conforms to minimal checklist requirements. Because the terms are used to describe many different programs with different requirements, the public cannot really understand the meaning of the terms and the programs without considerable research into the individual programs.

Registration of quality assurance programs and of manufacturers differ in the rigor of their requirements. Some registration programs register companies that pass assessments that include only product or process evaluation. Designers of registration systems must be careful in outlining the limitations of their registration and assessment so that consumers know exactly what assessments and registrations mean. In some cases, such as the FDA GMP system, registration entails an assessment of both product and process and therefore, this registration program is as rigorous as product certification programs that are described in Chapter Four. In other cases, the systems are unlike product certification, and are similar to the traditional grading of hotels signified by the number of stars — based on a simple assessment using a fixed checklist to rate a facility.

PUBLIC PERCEPTION AND CONFIDENCE OF PROGRAMS AND TERMS

> Went down and spoke at some lawyer's meeting last night. They didn't think much of my little squib about driving the shysters out of their profession. They seemed to ... doubt just who would have to leave.
>
> Will Rogers

This quote ridicules self-assessment — which is synomous to professional registration of one group of professionals. It could be expanded to many other professions. The public perception of existing accreditation, certification, listing, registration, and other evaluation programs can best be described as "confused" or mistrustful.

One EPA study[22] published an evaluation of the EPA's only certification program and stated "While these requirements provide some measure that the data generated are of a known and acceptable quality, the individual criteria vary among state to state certification programs. The periodic checks (PE samples and on-site inspections) for maintaining the certification are not designed to monitor laboratory ongoing performance." The study further provided an assessment of the

Agency's de facto certification for the analysis of waste samples. It stated:

> "However, since the controls in the CLP are only monitored for those laboratories under contract and providing analytical services to the Agency, there is no assurance that data generated by industry, or by CLP laboratories on non-CLP samples are of the same quality. Fully-qualified laboratories which might chose not to compete for purely business reasons, or were not selected for CLP contracts could be implied to be second class."[23]

Registration and licensing programs are usually not well understood by the public, partially because of the diverse nature of the requirements for different registration programs. Adding to this confusion are the different types of programs, such as product registrations, quality registrations, and registration of individuals programs. All of these systems have different requirements and bases. In addition, since licenses are sometimes a second layer of approval that cannot be awarded unless another system's requirements are met, the rules for licensing and what this term connotes about an entity is even less well understood.

It is not surprising that the public perception about the merits of the various evaluation programs is best described as "confused." The myriad of programs whose sole intent is to establish credibility for laboratory data fail to meet this goal. This is partially because of the profusion of different types of programs. Some of the programs do, however, instill public confidence in the quality of selected data. In the following chapters, a few successful programs will be examined for the purpose of establishing why the public has confidence in these programs and to provide a basis for design of a credible system.

REFERENCES

1. *Webster's 9th Collegiate Edition* (Springfield, MA.: Merriam Webster).
2. Whitney, Scott C., "The Case for Reforming the Environmental Protection Agency's Scientific Research Program and Establishing A Uniform National Laboratory Accreditation and Certification System", *Virginia Environmental Law Journal* 9(1): 134–142 (1989).
3. Dux, James, "Who Says Your Lab Data Is Good Enough? — Lab Accreditation Eliminates Questions About Data Quality," *Industrial Chemist*, pp. 36–40 (January 1988).
4. Chumas, Sophie J., Ed., *Directory of United States Standardization Activities*, National Bureau of Standards Special Publication 417, U.S. Government Printing Office, pp. 26 (November 1975).
5. Felix, Charles W., and George A. Kupfer, "Third Party Certification — A Measure of Protection for the Consumer," *Journal of Environmental Health* 50(6): 347–351 (1988).
6. Breden, Leslie H., "The Certification of Building Products in Europe," *ASTM Standardization News*, pp. 42–45 (June 1989).
7. Berman, Gerald A., "ASTM Committee E-36 Activities in Standards Development for Laboratory Evaluation and Accreditation" in *Evaluation and Accreditation of Inspection and Test Activities*, Harvey Schock, Ed., (Philadelphia, PA: ASTM Special Technical Publication 814, 1983), p. 11–17.
8. Berger, Harvey W., "The National Laboratory Accreditation Program (NVLAP) - Guest Editorial," *American Laboratory*, 18(11): 8–10 (1986).
9. Craig, David H., Carol J. Kelly, and John W. Locke, "Accredited Labs Can Build Confidence With Measurement Assurance Checks," *Quality Progress*, pp. 48-50 (May 1988).
10. "Position of the Chemical Manufacturers Association (CMA) on Laboratory Accreditation, and Quality Assurance and Control," Chemical Manufacturers Association (CMA) - (Approved 1988) paper presented at Consortium for Quality Environmental Data Meeting, Washington, D.C., (September 1989).
11. Carlberg, Kathleen A., "Enseco's Position on Laboratory Accreditation," Enseco, Inc. (May 1988).
12. Gray, Julian C., "Selecting a Lab — Not Just a Game of Chance," *Asbestos Issues'89* (August 1989).
13. Hess, Earl H., "Certification and Accreditation, *Environmental Lab* (July 1989).
14. "Availability, Adequacy, and Comparability of Testing Procedures for the Analysis of Pollutants Established Under Section 304(h) of the Federal Water Pollution Control Act, Report to the Committee on Public Works and Transportation of the House of Representatives, EPA/600/9-87/030, pp.6.3–6.5 (1988).
15. "Availability, Adequacy, and Comparability of Testing Procedures for the Analysis of Pollutants Established Under Section 304(h) of the Federal Water Pollution Control Act, Report to the Committee on Public Works and Transportation of the House of Representatives, EPA/600/9-87/030, pp. 6.4–6.5 (1988).
16. "American National Standard for Certification — Third-Party Certification Program," ANSI Z 341-1987 (American National Standards Institute).

17. Chumas, Sophie J., Ed., *Directory of United States Standardization Activities*, National Bureau of Standards Special Publication 417, U.S. Government Printing Office, p. 9 (November 1975).
18. Communication from the State of Washington to Licensed Professional Engineers, (Fall 1989).
19. "Good Manufacturing Practices for Medical Devices: General U.S. Code of Federal Regulations, Title 21, Part 820, (1978 and as amended 1988).
20. Willborn, Walter, "Registration of Quality Programs," *Quality Progress*, pp. 56-58 (September 1988).
21. Van Nuland, Yves, "The New Common Language for 12 Countries," *Quality Progress* (23)6: 40-41(1990).
22. "Availability, Adequacy, and Comparability of Testing Procedures for the Analysis of Pollutants Established Under Section 304(h) of the Federal Water Pollution Control Act, Report to the Committee on Public Works and Transportation of the House of Representatives, EPA/600/9-87/030, p. 6.4 (1988).
23. "Availability, Adequacy, and Comparability of Testing Procedures for the Analysis of Pollutants Established Under Section 304(h) of the Federal Water Pollution Control Act, Report to the Committee on Public Works and Transportation of the House of Representatives, EPA/600/9-87/030, p. 6.5 (1988).

Chapter 3

PRINCIPLES OF LABORATORY ACCREDITATION

DESCRIPTION

> "It is characteristic of science and progress that they continually spin new fields to our vision."
>
> L. Pasteur

The term "accreditation" is generally associated with attempts to provide evidence for capability to perform. Accrediting bodies seek to identify a minimum level of competence in performing a service or series of tasks or functions. It implies that the accredited entity (either an individual or organization) has been evaluated or tested by an authority widely recognized as having the expertise to accurately judge the capabilities of the accredited entity to perform a selected list of activities. Further, it implies a level of credibility and acceptance unobtainable without the seal of accreditation.

From a humble beginning of only two or three recognized laboratory accreditation systems in the 1940s, the number of formal systems currently operated in the United States is nearing 100. The majority of these systems are sponsored or operated by the federal government and by state and local governments. Approximately one-third of these formalized systems are activities of professional and trade associations. In addition to these formalized systems, a myriad of both formal and informal private accreditation systems exist to satisfy contractual requirements between purchasers and suppliers of goods and services .[1]

While nearly all current accreditation systems differ markedly in detail, there are many similarities that can be considered nearly generic in nature. Many listings of criteria required for laboratory proficiency testing programs have been written. One such standard is the American Society for Testing and Materials (ASTM) E1301-89.[2] Paramount among these requirements are the need for a competent, adequately trained staff, a qualified director and technical area managers, adequate facilities, appropriately maintained and calibrated equipment, and in most systems — a quality assurance function. Beyond these generic attributes, the specific details or criteria are the defining benchmarks for each accreditation system.

Regardless of the paucity or breadth of the specific criteria, every accreditation system must include a process for evaluation. These evaluations vary from extremely cursory examinations of the capability of the accredited entity to perform acceptably based on application forms to extremely detailed evaluations involving elements ranging from proficiency testing to complete data audits. In general, the more sophisticated assessments involve performance evaluations, on-site visits to ensure adequacy of facilities and equipment, review of technical personnel qualifications and training, determinations of adequacy of instrument maintenance and calibration, selected data review, and quality assurance audits. The object of this evaluation process is to ensure that the most important goals of any successful accreditation system — credibility and acceptance — are met.

At this point, it might be appropriate to ask if all the effort that is included in laboratory accreditation is really necessary. Consider for a moment the quote from Louis Pasteur presented at the beginning of this chapter:

> "It is characteristic of science and progress that they continually open new fields to our vision."

We are all citizens of a rapidly changing, technically complex world. Unlike the fabled "renaissance man" of past centuries, we cannot be expert in all fields of endeavor. Consequently, we must all accept — either on faith or with enlightened confidence — the results of efforts of highly trained specialists in a wide array of technical areas. It is certainly true that science and progress have opened new fields to our vision. In order that we can adequately understand our expanded field of vision requires an assurance that data are acceptable for their intended use. Meaningful accreditation of laboratories is not only desirable, it is essential in today's complex technical environment and mandatory for the future. Meaningful laboratory accreditation provides the assurance that the data are acceptable for their use, even if we, individually, cannot directly assess the quality of the data.

STRENGTHS AND WEAKNESSES

Accepting the premise that laboratory accreditation is not an evil to be endured, but is rather, an extremely desirable and worthwhile goal, a key question is raised. Several similar questions were raised in Chapter 1. In Chapter 1, the credibility of the current accreditation and certification programs was questioned based on decision maker's acceptance, trust, and use of the programs. This concept is repeated here by posing a slightly different question. It is repeated because the answer to this question is so vital to understanding why the current accreditation and certification schemes cannot, as designed, garner acceptance, trust, and use. This key question is, Are current systems providing the seal of credibility attached to the accreditation concept?

While it is beyond the scope of this discussion to address the pertinent aspects of the numerous accreditation systems in existence, several generalizations are offered in response to the question posed.

The key question is best answered by considering four separate questions. These are:

1. Are the accreditation and certification systems functioning appropriately?
2. Are the accreditation and certification systems' design providing continual monitoring of laboratory work?
3. Are data purchasers accepting the systems as capable of providing a measure of laboratory quality?
4. Are the systems providing the seal of credibility that is associated with the terms accreditation and certification?

Current accreditation systems range from rather limited evaluations to stringent programs. The former are rather informal and have criteria that consist of a few contractually based requirements. The latter systems include rigorous requirements and include continual oversight, such as the EPA's de facto certification program — the Contract Laboratory Program.[3] Therefore, since there are so many different systems, it is difficult to determine if, in general, systems are functioning. In addition, many systems might not be functioning well, but might have been initially designed to provide a more detailed assessment than other programs. In such cases, these systems could be providing a better indication of laboratory quality, even if not functioning well, than other programs that are less rigorous in their initial design but are functioning well. Therefore, one must consider both the rigor of the certification and accreditation design, and also whether the system is working well as designed.

Without a doubt, every accreditation entity has some confidence in their particular system. What is questionable, however, is the overall effectiveness of these systems. With few exceptions, most notably the EPA's Contract Laboratory Program, the current accreditation systems represent snap-shot evaluations of a laboratory's capability to perform the accredited tasks. Most accreditation and certification programs assume continued quality performance of the accredited laboratory during the interval between evaluations. While nearly all laboratories want to do high quality work, the requirements present in essentially all current accrediting systems do not *assure* the continuance of acceptable performance.

Nearly all current accrediting systems have limited monitoring capabilities. Rather, the normal evaluation process is designed to determine the capability of laboratories to perform the desired task or service. Even systems as rigorous and thorough as the EPA's Contract Laboratory Program[4] only evaluate the laboratory's ability to perform on actual blind samples on a quarterly basis. Moreover, the currently used performance evaluation materials seldom resemble environmental samples. Therefore, the materials seldom truly assess laboratories' capability to perform on the types of matrices of interest to the data purchasers.

Current accreditations are not widely accepted by purchasers of laboratory services. Limited reciprocity or acceptance exist between the various accreditation systems. This lack of wide acceptance results in duplicative effort that is extremely costly in terms of both time and money. "Political reasons" are often cited to explain this unfortunate situation. However, the fact remains, that at this point, there is no single national accreditation or certification program for environmental laboratories that is acceptable to all purchasers of data. One key reason for this is that a single, nationally accepted set of accreditation criteria does not exist. Until a widely accepted, truly independent and technically competent accreditation entity exists that is based on the accepted set of criteria, the many current systems will be in conflict and will result in unnecessary duplication of effort. Until the government assists in developing uniform criteria for accreditation and certification, the development of such criteria will be stymied. Until the government accepts a third party system as a part of the government's system for evaluating laboratories and their products, a uniform third-party system will not exist.

It is apparent from the previous discussions that current systems are not providing the seal of credibility attached to the accreditation concept. However, within limits, this question has a positive answer. The majority of accreditation systems provide some indication of the capability to perform laboratory services in an acceptable manner. As such, these systems provide a starting point from which to design a better system that will garner acceptance, trust, and use.

PUBLIC PERCEPTION AND CONFIDENCE AND SCIENTIFIC VALIDITY

There are several widely recognized and generally accepted accreditation and certification criteria for products or individuals offering specialized services that currently exist in the United States. Among these are the American Medical Association for medical practitioners, the criteria for obtaining the designation of Certified Public Accountant and the product approval process provided by the Underwriter's Laboratories. There is no laboratory accreditation program that is widely accepted in this country. This is partially because a uniform and accepted set of assessment criteria for evaluation of laboratories does not exist.

Several of the more widely recognized attempts at defining benchmark criteria are

1. American Society for Testing and Materials (ASTM) E548 — Practice for Generic Criteria for use in the Evaluation of Testing and Inspection Agencies[5]
2. International Organization for Standardization (ISO) Guide 25[6]
3. Organization for Economic Cooperation and Development (OECD) Principles of Good Laboratory Practice for Testing Chemicals[7]
4. The accreditation program sponsored by the American Council of Independent Laboratories — the American Association for Laboratory Accreditation (A2LA)[8]

Obviously, no attempt has been made to enumerate every group or organization that defines accreditation criteria. No slight is intended or implied. The examples are selected from the existing sources to provide examples of criteria that have been used for developing benchmark criteria. The important point is this: "How widely recognized are these criteria by the community of laboratories in the United States?" Perhaps even more importantly: "Does the public understand the criteria associated with the accreditation process?" It is questionable whether the majority of laboratories in the United States are fully aware of the criteria promulgated by the above organizations. There is no doubt that the general public is unaware of, or not concerned, with these or any other laboratory accreditation organizations and the various criteria used.

Does this mean that all accreditation systems are to be forever transparent to the general public? Probably not. If we consider, as an example, the Underwriters' Laboratories (UL), it seems completely feasible that laboratory accreditation can gain wide public acceptance. In the case of UL, the public might not fully understand the nature or extent of the testing criteria required to obtain the UL approval, but they do understand that this approval indicates safety and reliability of the product. So trusted is the UL name that the majority of insurance underwriters in the United States and many Federal, state, and municipal authorities and other users either accept or require listing or classification by UL as a condition of their recognition of devices, systems, and materials having a bearing upon life and fire hazards, and upon theft and accident prevention.[9]

This acceptance by the public of the UL approval arises from the national scope of the UL program, their independence, their technical competence, and most importantly, the credibility associated with the UL approval. Currently there exists no comparable program for laboratory accreditation. Any such program has a number of obstacles to overcome. Among these obstacles are public perceptions of competence of accreditation and certification agent(s), independence of accreditation and certification bodies from the laboratory industry, accountability to maintain continual monitoring of quality, and the general lack of technical knowledge needed to understand the issues surrounding laboratory quality.

A workable approach to providing an accreditation system that is both widely accepted by the public and the laboratory community and has a high degree of credibility might be to emulate the UL approach. This means that a technically competent, independent third-party (i.e., one that is neither a purchaser or supplier of laboratory services) evaluates the performance of accredited laboratories and is accountable for maintaining a monitoring effort to ensure continued acceptable performance.

A key part of the approach is to employ a third-party for the accreditation. The critical reason that a third party must perform the accreditation is that they are independent. Therefore, they can assess performance against requirements in an unbiased manner. Further, the third party would be responsible for accreditation as their primary concern, not as a secondary part of their operations. For this reason, the accreditation authority could be technically competent. Third, the third party would be continually accountable to all programs, clients, laboratories and data users to which it provides accreditation services. The third party would be continually challenged by laboratories that were not pleased by assessments and likewise would be challenged by purchasers of data from laboratories if the data did not meet the implied requirements of the certification and accreditation. The third party would be continually challenged and accountable to provide evidence to all these parties that a competent and professional assessment of laboratories was conducted. An approach that uses a third-party will ensure wide acceptance and credibility and will engender a positive public perception.

COST AND BENEFITS

To reiterate the problems — the current accreditation systems are subject to politics, are not truly independent, are not widely accepted, are duplicative in nature, and generally lack the perception of credibility. The cost of such duplication of efforts to the laboratory community is high and is passed on to the consumer. In this vein, the cost seems to outweigh the benefits. Indeed, individual laboratories often spend tens of thousands of dollars yearly, merely to obtain the mandated accreditation.[10]

Consolidation of the various accreditation systems seems both unlikely — due to political considerations, and impractical — due to the diversity of specific criteria associated with the wide array of laboratory types (e.g., clinical, environmental, government, industrial, etc.). Recognizing the futility of attempting to force a single system of accreditation on the laboratory community leads us again to the concept of an independent third-party approach. Such an organization, if structured correctly, could develop and implement a baseline, generic assessment approach. This baseline assessment would reduce cost, in terms of both time and money, through the elimination of duplicate effort. It would also provide a more consistent evaluation of laboratories and overall monitoring of laboratory performance, increased accountability to the purchasers of laboratory services, and a more widespread acceptance and greatly increased credibility of the accreditation system and the laboratories alike. It would also allow the more specific assessments of laboratory performance to be measured by additional criteria and assessments that could also be conducted by a third party. In addition, it can free up resources to allow a good job to be completed in identifying and in assessing the additional criteria that are critical to government and industry that are not included in the national baseline, generic system. The resources that now are spent in redundant, baseline inspections can be used to provide real-time monitoring of critical elements that are specific to the projects for which the data are generated.

While this approach certainly is not an easy task, there are successful systems that exist that could be used as patterns. The National Sanitation Foundation's (NSF)[11] certification program for bottled water and the drinking water additives program are excellent examples of independent third-party oversight programs. The drinking water additives program was developed under formal agreement with the U.S. Environmental Protection Agency, and in cooperation with the Association of State Drinking Water Administrators (ASDWA), the American Water Works Association Research Foundation (AWWARF), Conference of State Health and Environmental Managers (COSHEM) and the American Water Works Association (AWWA). The standards that were developed in the additives program have been accepted in the marketplace through advisory opinions provided by EPA.

The initial hurdle to overcome to implement a similar program for laboratory accreditation is the reluctance of purchasers (federal, state, and local governments, the regulated industrial community, etc.) to accept the concept of an independent third-party accreditation system. The technical expertise and management acumen exist in all fields of laboratory services to successfully operate such a system. NSF has already expressed interest in assisting in the development of a program, if the EPA laboratories, appropriate trade associations, and data user clients can work cooperatively.[12] The appropriate trade associations have begun to discuss the possibility of working together on this effort.[13] The government authorities and other purchasers of data that will potentially use the system need to express interest in a cooperative effort for progress to be made.

EFFECTIVENESS AND INTEGRITY

The subject of the effectiveness of the current accreditation systems has been mentioned previously. Again, the sheer number of systems, the limited reciprocity, the nominal one-time evaluation, and the perception of limited technical competence within the accrediting organizations exacerbates the lack of effectiveness of the current programs. No matter how effective any given program is in its limited scope of application, the inherent limitation of its implementation inexorably limits the particular program's effectiveness. Even those programs that profess to be national in scope suffer the malaise of ineffectiveness due either to the limited acceptance or to inadequate oversight of laboratory performance. Even if mandated, these systems are often viewed as technically meaningless by the regulators and the regulated.[14]

This concern has been addressed by J.A. Cotruvo, Director of the Criteria and Standards Division, Office of Drinking Water, U.S. EPA, A. H. Perler, Chief of the Science and Technology Branch, Office of Drinking Water, U.S. EPA; and B.L. Bathija, Chief of the Monitoring and Exposure Section of the Office of Drinking Water U.S. EPA.[15] They have discussed the EPA's certification program for the laboratories performing water analyses established in 1978. and have stated that:

> The current EPA laboratory certification program (LCP) evaluates laboratory quality with an on-site inspection at least once every 3 years and an annual performance (PE) sample for every analyte for which the laboratory is certified.
>
> Experience has shown that there are a number of shortcomings in this certification process. The following are some of the major issues in the program, along with some possible options:
>
> • Performance on the PE samples could be an indication of how well the laboratory is capable of performing under the best circumstances and not how well it performs on a routine basis. It is a measure of how well a laboratory performs at a particular time, not necessarily now well it routinely performs.

- A system needs to be developed for assuring that the data produced on a daily basis are of known and verifiable quality. For example, it might be possible for the laboratories to document the quality of daily analytical results as they are generated by having the results and attendant precision and accuracy data reported together.

- Currently, a laboratory is required to satisfactorily analyze each analyte for which it is certified. With the increase in the number of regulated analytes, the laboratories will spend increasingly large portions of their time performing PE analyses. If a method can be used for many analytes, it might not be necessary for the laboratory to prove that it can analyze each and every one of them. For example, if a laboratory can satisfactorily analyze for a few selected analytes, it can often be assumed that the laboratory can also satisfactorily analyze for the others using the same method. Accordingly, future PE samples might contain only selected or representative analytes instead of all analytes. A laboratory could then be certified for a particular method rather than on an analyte-by-analyte basis.

- With the increasing number of drinking water contaminants being regulated, the probability of a good laboratory failing to analyze an analyte satisfactorily will increase because of the statistical variation of results from a large number of analyses. This was taken into consideration when laboratory approval criteria for volatile organic chemicals (VOCs) were considered. For VOCs, a laboratory must satisfactorily analyze for six out of seven regulated VOCs to be approved for all seven VOCs. This was reasonable because of the similar characteristics of the VOCs and the unlikelihood of matrix interferences. In the future, similar considerations could be made for a group of related compounds. The laboratories could be allowed acceptable failure rates and still remain fully certified.

- EPA regulations require that analyses must be performed in laboratories certified by the state in which they reside. A laboratory wishing to provide analytical services in more than one state has to be certified by all the states in which it provides services. Although the EPA strongly supports the concept of reciprocity between the states, there is little reciprocal certification granted by the states because they are reluctant to

> accept one another's certification unless their criteria are equivalent.

The authors conclude:

> There are a number of certification/accreditation programs currently functioning in the United States. This gives rise to duplication of effort and waste of resources. Many certifications and pseudo-accreditation are produced, none of which is inappropriate in itself; but collectively they sap the energies of a testing laboratory. The obvious solution would be a national comprehensive, generic program that successfully preserves the two essential values of technical soundness and practical efficiency. The underlying philosophy of this national system would be to unify, not fragment; to be efficient, not duplicate; and to be demanding, but reasonable. Its approach would be to start with a broad field of testing and to move from it to narrower testing technologies within that field.

It is heartening to have such influential people within a regulatory agency espouse the avenue of reason and not the politically safe avenue of self-interest. It is to be hoped that such an enlightened view will become common throughout the various regulating bodies, and that a system such as briefly described by the authors from the EPA's Office of Drinking Water will become a reality — a reality based on an independent third-party laboratory accreditation system.

REFERENCES

1. "Principal Aspects of the U.S. Laboratory Accreditation Systems", U.S. Department of Commerce, NTIS Acquisition Number PB80-199-86, National Technical Information Service, Springfield, VA.
2. "Standard Guide for Development and Operation of Laboratory Proficiency Testing Programs," ASTM E 1301-89 (Philadelphia, PA: American Society for Testing and Materials, 1989).
3. "Comparison of the New York Environmental Laboratory Approval Program (ELAP) With Other Quality Assurance Programs," Report to the New York State Division of the Budget prepared by Dynamac Corporation, 1988, Part 3.
4. "USEPA Contract Laboratory Program Statement of Work for Organics Analysis, Multi-Media, Multi-Concentration, SOW No. 2/88, including Rev.9/88 and 4/89, p. E-67.

5. "Standard Practice for Preparation of Criteria for Use in the Evaluation of Testing Laboratories and Inspection Bodies," American Society for Testing and Materials (ASTM) E 548-84 (1984).

6. "General Requirements for the Technical Competence of Testing Laboratories," International Organization for Standardization (ISO) ISO/IEC Guide 25 (1982).

7. Friedman, Ira J., "The Importance of the Protocol," Managing Conduct and Quality of Toxicology Studies in *Principles of Good Laboratory Practice for Testing Chemicals*, B. Kristin Hoover, Judith K Baldwin, and Arthur F. Uelner, Eds. (Princeton, NJ: Princeton Scientific Publishing Co. Inc., 1986) pp. 119-124.

8. Locke, John, "Developing an Environmental Laboratory Accreditation System," paper presented at a meeting of the A2LA Environmental Advisory Committee (July 7, 1987).

9. Chumas, Sophie J., Ed., *Directory of United States Standardization Activities*, National Bureau of Standards Special Publication 417, U.S. Government Printing Office, p. 162 (November 1975).

10. Farrell, John, "Do Multiple Certification and Accreditation Programs Enhance the Quality of Environmental Data Generated by Commercial Laboratories - An Alternate Approach" presented at the USATHAMA Quality Assurance In Environmental Measurements Meeting, Las Vegas, NV (May 1989).

11. McClelland, Nina I., "The National Sanitation Foundation's (NSF's) Third-Party Programs to Assist States," presented to the Association of State Drinking Water Administrators (ASDWA), Tampa FL (February 1989) p. 4.

12. McClelland, Nina I., "The National Sanitation Foundation's (NSF's) Third-Party Programs to Assist States," presented to the Association of State Drinking Water Administrators (ASDWA), Tampa FL (February 1989) p. 8.

13. "Accreditation Committee Update," *Environmental Testing ADVOCATE*. II(1): 1,4 (1990).

14. "Availability, Adequacy, and Comparability of Testing Procedures for the Analysis of Pollutants Established Under Section 304(h) of the Federal Water Pollution Control Act, Report to the Committee on Public Works and Transportation of the House of Representatives, EPA/600/9-87/030, p. 1-8 (1988).

15. Cotruvo, J.A., A.H. Perler, and B.L. Bathija, "Certification and Accreditation," *Environmental Lab* 1(4): 38–39 (1989).

Chapter 4

PRINCIPLES OF PRODUCT CERTIFICATION — AS RELATED TO LABORATORY ACCREDITATION

DESCRIPTION

The previous chapter described laboratory accreditation. "Good" laboratory accreditation programs inspect facilities and otherwise determine if a laboratory is capable of producing acceptable quality data. Less rigorous laboratory accreditation programs use less rigorous ways to attempt to determine if a laboratory is capable of producing acceptable data. This chapter will describe product certification. A good product certification program does not determine capability to produce. A good product certification program determines if the process is capable of producing acceptable products and then goes one step further; the product certification program inspects at least one product that is produced by the inspected process. The difference between laboratory accreditation and product certification is that product certification is more rigorous.

In addition to information about product certification, this chapter includes information about some common product registration, registration of individuals, registration of quality programs, and product and service "listing" programs. These programs have been included in order to show the similarities and differences between these programs and product certification programs. They have also been included to provide insight into the difficulties encountered in attempting to categorize and define the rigor of programs because of semantics.

Before the third-party approach and programs are described, it is useful to fully define a third-party as contrasted with other "parties." The American National Standard for Certification — Third-Party Certification Program ANSI Z-34.1-1987 states in Section 4. of the standard[1] that:

The certification body whose name is identified with the program shall be one of the following:

1.   A trade association
2.   A professional or technical society
3.   An organization of producers of service-rendering entities
4.   An organization oriented to consumers or users of the product or service
5.   A third-party testing/inspection organization

Dr. Nina I. McClelland, President of the National Sanitation Foundation has provided[2] an explanation of what the term third-party means to classical third-party organizations like NSF. First-party certification is self-certification by the manufacturer or producer. Second-party certification is conducted by a private, independent company, a trade association, or another agent of the producer on a direct fee for service basis. First-party and second-party certification typically involve a direct relationship with the producer of goods and services, so therefore are not independent of the producers. A third-party certification is one that serves the needs of one or more parties other than the producers of goods and services. NSF and UL are this type of third-party. Fourth-party certification is conducted by an official agency such as U.S. EPA. In this case, diverse public concerns are served, as well as the producer's. Third-party and fourth-party certifications have more than a client-only relationship.

A few well known product certification programs are now briefly described and the principles of this type of system are summarized and related to laboratory accreditation.

THIRD-PARTY PRODUCT CERTIFICATION PROGRAMS

The third-party concept was begun many years ago in 1893 in Chicago. The incandescent electric light bulb had been invented for only 14 years at that time, and there was still much that was unknown about the art of wiring and electrical hookups. William Henry Merril, an early electrical investigator, was employed at the Palace of Electricity at the great Columbian Exposition in Chicago to solve the numerous problems caused by the tangle of wires that were constantly starting fires and threatened to raze the entire Exposition. During his trials in attempting to create safe wiring out of the tangle of wires, the idea to begin a third-party certification program for electrical items occurred to him. A year later, he founded the Underwriters Laboratories.[3]

Underwriter Laboratories (UL) became perhaps the best known product certification program.[4] A consumer that is looking for assurance that an electrical appliance has been tested against an official or accepted standard for safety or performance and found to be acceptable looks for the UL mark. This mark allays a purchaser's concerns that a appliance that is purchased will cause reasonably foreseeable hazards to life and property.

The National Sanitation Foundation (NSF) mark instills similar confidence in the public health area. The NSF mark signifies that food service equipment that bears the mark conforms to the materials, design, construction, and performance that could otherwise contribute to cross contamination of foods in restaurants. An NSF mark can also be found on bottled water, packaged ice, plastic pipe, and many other products. This mark signifies that the product conforms to specific standards of safety and performance.[5] Recent articles on NSF have identified NSF as a key player in efforts that industry and government are making to improve drinking water in the United States.[6]

What do these two product certification systems have in common? If the reader is familiar with the systems, the most obvious similarity is that they are implemented by third parties. Is this the critical feature of the programs that makes them so credible? No. This is a necessary attribute of the implementation, but the critical aspect of the highly credible programs is one distinct feature — the third-parties that implement these programs not only develop standards, but also perform follow-up inspections. These distinctly different product certification programs complete the standards writing process with a program of testing and certifying the conformance of products and services to the standards. It is this act of rigorous inspection of products and facilities to check on the continued compliance in the production of products that sets these organizations apart from other third-parties. After inspection and certification of the products, the UL or NSF mark is attached to the products. This mark signifies that the third-party completed all necessary inspections, in accordance with the appropriate standard, and that the articles produced by the manufacturer met the requirements.

It is important, at this point, to differentiate between development of standards and affixing a third-party mark to items. Many third parties develop standards.[7] Further, the standards developed by UL and NSF can be used by other third parties. However, the UL mark and NSF mark, as well as any other like mark that is registered to a specific third-party organization must not be used by any other entity.[8]

Underwriters Laboratories was the first third-party program.[9] Since its inception, countless other third parties have sprung up. Most of these programs have originated to develop voluntary consensus standards for products that are acceptable to both industry and the regulators. Many of these standards writing programs have been merged into the American National Standards Institute (ANSI),[10] the largest standards writing organization in the United States. This organization has processed and adopted over 8500 standards for a wide array of products and services since it began its process in 1918. In the international arena, the International Organization for Standardization (ISO) serves a similar function in the international arena.[11] ANSI is a member of ISO. Third parties usually only develop third party consensus standards. The feature that differentiates UL and NSF from the other third parties is that they actually test products to determine if the products conform to the consensus standards.

LISTING PROGRAMS

Product listing is not a well understood concept. This is partially because it can be considered one type of inspection of product to requirements, that is, a type of product certification. Good examples of the term listing are the NSF Listing for certain products and the UL Listing and Inspection Service. NSF uses both NSF Listed and NSF Certified to differentiate between two slightly different inspection and verification processes. Certification is used to denote that a product has been tested for conformance to regulations or a standard that was not developed by NSF (such as a standard developed by ASTM or ANSI). Listing is used to denote that a product has been tested for conformance to NSF consensus standards. UL has several programs in addition to its Listing program. Each separate program connotes a slightly different evaluation system with different requirements. However, the terms listing and product certification are almost synonymous in NSF programs and the terms are also almost equal in the UL programs.

STRENGTHS AND WEAKNESSES

Product Certification Programs

Product certification has many strengths. The major product certification entities that have been described here inspect both processes and products for compliance to specific standard criteria. Instead of inspecting only factories' capabilities to produce quality products, the products are inspected to show that the factories produce quality products. The factories are usually inspected on a periodic, unannounced basis. While at the factories, the inspectors check to determine if the factories are routinely operated according to the standards criteria in the production of products that bear the third-party certification mark. In addition, if the products cannot be adequately tested during the on-site inspection, products are routinely laboratory tested by the third-party for conformance to the product standard criteria.

Another strength of product certification is that the cost of the certification is borne by the industry that is served by the certification service and the users of the industry's service. It makes more sense for consumers and taxpayers to have programs administered by a third-party certification service when possible. A third party, rather than the government, conducts the inspections, and the industry and the persons that benefit from the service are responsible for the cost involved.

One clear strength of product certification is that the mark of approval of a competent third party on a product is valuable to the consumer. A consumer can easily differentiate between a product with the certification and a product that bears no third-party mark of conformance. Consumers can identify which products are made according to accepted standards and which products are of questionable quality. This is a great benefit to consumers. This is also a great benefit to the industry that produces the products. The industry is protected, in part, from manufacturers that produce shoddy products. Consumers can use the certification mark on quality products to differentiate between quality products and suspect products. This, in turn, discourages manufacturers from producing inferior products because the market for them is limited since they cannot pass standards criteria and bear the certification mark.

Also, these inferior products cannot command a high price or compete with the goods that are manufactured according to strict standards. Consumers can readily tell the difference in quality of different products from the certification marks.

One weakness of third-party certification of products is the reluctance by some individuals in the government, industry, and with some consumer groups to accept third-party certification. This is not a weakness of the system, but a lack of confidence due to incomplete understanding of the system by these individuals.

Another weakness is the confusion between real third-party product certification and other "certification" programs. The use of the term certification in the context of "laboratory certification" and "board certified" causes confusion about the true meaning of certification. The "certification programs" that do not perform product inspection have the term because it connotes credibility. The use of the term for anything but actual product certification dilutes its value to be immediately recognized as denoting product conformance to a standard. Again, this weakness is not with product certification, but with the use of the term for other purposes.

The most critical weakness with product certification is that, as with anything that is successful, many entities could attempt to develop and implement third-party certification programs. The quality third-party certifiers are recognized by their reputations and well known certification marks. The lesser known third parties do not necessarily produce poor work, but also do not necessarily implement good programs. Consumers, industries, and regulatory officials alike should monitor their efforts closely — because the lesser known third parties must prove that they are credible. The third-party field is, to some extent, self-policed, because failure of the certifications are evident in most cases. In addition, the regulators role in third-party product and service certification is to monitor the activities of the third parties that implement programs under their purview. This monitoring function is not as resource intensive as administering their own certification programs, but is essential for the regulators to maintain control of the process.

Essentially, third-party certification of products has far more strengths than weaknesses. The weaknesses that it does have can be dealt with and corrected or contained.

Listing Programs

Product listing programs have the same strengths as product certification and registration. However, the strength of the programs could be diluted by the lack of understanding the term "listing." This term might not be understood to represent the same rigorous process as product certification.

PUBLIC PERCEPTION AND CONFIDENCE AND SCIENTIFIC VALIDITY

Product Certification Programs

Why do people trust the certification marks from these credible third parties? What makes them credible? Charles W. Felix described what he termed the "six Cs of Success" in an article entitled "Third Party Certification."[12] The public has confidence in the integrity of the certification programs implemented by third parties as a direct consequence of the six Cs. Reputable third-parties have made these attributes integral to their organizations. Felix stated that the six C's are: competence, credibility, confidence, confidentiality, communication, and consensus. The importance of these attributes is self-evident. However, a brief description of how the terms relate to what happens when a third party succeeds in both development of standards and subsequent implementation for certification of products and services stresses their importance. The six C's are described here in detail, as Felix explained their importance in his article.

The first C stands for competence. The need for competent people both in the technical and managerial areas is mandatory in order to manage the process of developing a consensus standard. A competent technical staff is necessary to assist in developing standards that are technically feasible to implement. Third parties that also implement these standards such as UL and NSF have a keen eye leveled on the feasibility aspect of the standards process from day one. The managerial staff must be able to move a consensus process efficiently to the final product. This process of building consensus among many contrasting positions requires competent and diligent management. The attainment of consensus without diluting the rigor of a technical standard is a real challenge. However, if the standard is so weak that any product will conform to the requirements, there is no need for the standard, and credibility in the certification mark of the third party would be diminished. The third party does have an intrinsic need to develop consensus standards that are credible.

According to the journal article, the second C, credibility, means that the third parties have neutrality that is essential to the success of certification programs. Third parties avoid even the appearance of complicity with clients that seek to have their products certified. The two product certification programs that have been described in this book are both independent, not-for-profit corporations. They have no vested interest in the products or services that they certify. They make periodic unannounced visits to factories to check on the compliance of the production of the products that are certified. The certification mark is revoked from use on products that do not comply.

The third C, confidence, is instilled in the entire standards building process by a credible third party. Each of the three parties that develop the standard in the standards development process must have confidence in the third party that is facilitating the development process. This confidence is a key feature in assuring success of the endeavor. It is because people from industry trust the third party to be neutral and competent that they are willing to put aside their suspicions about competitors and work with them to develop the standards. They know that there will be no collusion to develop a standard that is biased toward one manufacturer or another. The regulators also trust a reliable third party to assist in developing a standard that is based on true consensus. They have confidence that the standard will reflect their views because the third party gains nothing if the standard is not acceptable to the regulators and the regulated parties alike. The regulators would simply ignore the standard if it did not meet their needs. The users of the products would similarly not be interested in compliance testing to a standard that did not include requirements for testing that would increase their confidence in the quality of products. Manufacturers would not submit their products or services to testing in accordance with a standard that does not include their product quality concerns. If the third party strives to keep its reputation of developing credible and technically valid standards, they have good reason to carefully consider all user concerns and include them in the final consensus document.

The fourth C stands for confidentiality. Third parties must have the confidence of industry that trade secrets, new products, manufacturing methods and other proprietary information to which they have access is kept strictly confidential. Third parties have access to such confidential information in two different ways. First, they test products in the third-party laboratories. Sometimes these tests are completed on prototype products to determine if they will meet the standards' requirements before full scale production is begun. The third party certifying body also gains access to trade secrets through its periodic unannounced plant inspections. In the Journal of Environmental Health, Dr. Nina I. McClelland,[13] of NSF states in direct terms, "We wouldn't be in business for very long if we didn't know how to protect the proprietary interest of our clients."

Communication is the fifth C. Communication is necessary in the standards development process in order to facilitate the open dialogue between all three parties in the standards development process. The third party also communicates to the public the meaning of the standards that are developed. The standards are published in order that the public clearly understands what the requirements of the standards are and what it means for a product to comply with the standards. Third parties educate the public about what their certifications mean and publish listings of products that comply with standards. Third parties communicate their certification activities to the industries, usually through trade associations.

The sixth C, consensus, concerns the very thing that stands in the way of most industries, regulators, and users of products and services and limits their ability to develop compliance requirements. One critical function of the independent third-party is that they bring all three parties together to reach consensus. How can they achieve this when it is usually difficult to bring several competing companies together to agree on issues? Also, an even greater hurdle might be to effect consensus among many competing agencies of government, for instance health versus commerce and environment versus agriculture. State versus federal versus local can also be considered competing organizations in many cases when their requirements differ. Third parties that specialize in standards development are skilled in the art of achieving general agreement among all parties involved in the standards development. This is the major thrust of their business and they have a great deal of experience in this area. Many of the contributors to the consensus standard do not have time to make the standard their first order of business. They will not trust a standard that is developed by another competing agency. They will certainly not accept a standard that industry takes the lead in developing, even if they are on the periphery of the process. Clearly, they wish to develop a standard but do not have the time to control the process. Therefore, they feel uncomfortable about standards that are produced by others. The third party that is neutral to all players is the solution to collective problems of all parties that want a standard but either do not have the time to develop it themselves or trust others to develop a standard that will meet their needs. The key to consensus that the third party can achieve is that all parties will agree on a solution if a competent, neutral third party can effect the communication, technical insight, and expend the resources to make it all happen.

Listing Programs

Product listing programs are either understood by the public to be synonymous with product certification programs or not understood at all. In any event, the terms UL Listed or NSF Listed are recognized as meaningful if for no other reason than because the organization that is associated with the "listing" is credible.

COST AND BENEFITS

Product Certification Programs

The cost and benefit analysis of the use of third parties to develop standards is clear to both the government and industry if they are informed about the subject or if they have first-hand experience with the process. One government official[14] that understands the role of third parties has clearly stated the benefit of the use of third parties. Arthur L. Banks, Chief of the Retail Food Protection Branch of the U.S. Food and Drug Administration has stated:

> Increasingly governmental agencies at all levels of government are not able to do that (gain consensus to develop standards)...The laws are there and regulations are there, but they can't do it. They simply don't have the resources, and they can't make everything a high priority... When you get an industry working hard on public health and safely objectives, individually as companies or collectively through a trade association, you can accomplish more in a couple of years than you can in twenty years relying entirely on the regulatory process.

The U.S. EPA decided to take the third-party route to develop standards and to test and certify additives for drinking water.[15] The decision was based, in part, on the Gramm-Rudman-Hollings bill's influence on the national budget. There were simply not enough resources to develop standards in the Federal sector.[16] One particularly interesting aspect about production of third-party standards is that the industry — even though it has only part of the final vote, pays the bill for the standards production. The industry, therefore tries to communicate effectively with the other members of the standards production group because they are paying the bill. All standards production by NSF have historically been funded by industry. Third-party organizations can collect resources from industry and accomplish the production of a standard without burdening the federal budget. In the past, it has been the industry and not the taxpayer that has paid the bills in NSF third party programs.

It is interesting to note in this cost and benefit analysis that the government and the consumer groups have not been burdened with any of the standards development cost even though they are beneficiaries of the product standards. Why aren't these two groups required to pay a portion of the development cost? The consumer groups likely do not have adequate funding bases, but might be able to initiate and partially fund standards development in areas that are of critical interest to them. In many cases, the consumer groups lobby to have the regulators produce legislation to address their critical causes. The government is then accountable to produce implementable legislation to alleviate the consumer group's concerns. In many cases, product standards and a means to inspect products for compliance to the standards is the solution. If the government elects to use the services of a credible third-party, perhaps they should assist the effort by partially funding the third-party process? This is not how it currently works. If government co-funding were available, it could considerably shorten standards development time, because more resources to push projects to completion would be available. In the future, it might be reasonable to investigate the possibility of government paying an appropriate share of the cost, perhaps for speeding the process of initial standards development and for the maintenance of the previously developed standards to make sure that they keep abreast with rapidly changing technology.

Listing Programs

The cost and benefit analysis for product listing programs is the same as that which was provided for product certification programs.

EFFECTIVENESS AND INTEGRITY

Product Certification Programs

Third-party certification programs that are credible, such as NSF and UL, are effective and are trusted by the public, government, and industry alike. They are prime examples of the correct implementation of programs that differentiate poor quality products from acceptable quality products and services. This differentiation and the public's trust in the programs that accomplished the differentiation is the purpose of accreditation and certification. Third parties such as NSF and UL have mastered both objectives. Their programs are both effective and have integrity.

Are all third party services effective and trusted? It depends on the strength and adherence to the "six Cs" of a third-party organization. The universal trust that the reputable third parties possess has not been achieved by marketing and public relations — it has been gained through strict, long-term adherence to the "six Cs." These qualities might not be integral to every third party. Some third parties are established to perform a narrow field of inspections and do not have a track record. They might be worthy of trust and could be adhering to the "six Cs." One should determine if they are, in fact, reputable before accepting their standards and certifications.

In a recent presentation,[17] Dr. Nina I. McClelland, of NSF, described the elements that can be applied by government entities and other users to determine equivalence with NSF. The reason for this check of equivalence is to determine if other third-party programs are equivalent to the NSF third-party program. These elements can be used to determine if a third party has the necessary attributes to perform well. The elements that were listed by Dr. McClelland are directly quoted:

1. Fiscal and operational controls independent of producers of products/services being certified or their trade representatives
2. Fiscal integrity sufficient to ensure that the gain or loss of a specific client or program will not significantly impact its future viability
3. Well developed, clearly stated policies, procedures, and contracts to support enforcement procedures for meeting compliance objectives
4. An administrative infrastructure with legal support to effectively meet contractual commitments

5. An established system for investigating complaints and taking appropriate action, with an effective appeals process in place

6. One or more formally registered Marks, used to indicate certification of products (or materials) and services; without a registered Mark, regulators, users, and the third party may find themselves without recourse when addressing noncompliance

7. Policies and procedures for in-plant audits at reasonable, but regular frequencies to select samples for testing, evaluate quality assurance and quality control (QA/QC) procedures, review purchasing and shipping records to assure that only accepted ingredients are used, and observe production operations

8. Procedures for sampling from the field or marketplace

9. Established policies for periodic retesting and/or reevaluation; it is not appropriate for the only source of data and QA/QC documentation to be the producer or his retained agent; a third party is responsible and accountable for control of product quality and use of the certification Mark

10. Facilities and instrumentation adequate and appropriate for performing testing required by the standard and relevant certification policies

11. Qualified, competent staff to perform tests, make informed decisions, and to properly manage any and all subcontractors

12. Effective, non-conflict liaison with regulatory, code, and user groups served.

A reputable company such as UL will not enter into a certification program unless it will enhance, or at least not damage, the company's reputation. Likewise, NSF will not develop a program to certify a product to a standard that is not acceptable to industry and regulators. It is simply not in the best interest of these companies to do anything to taint their reputation. These third-party certifiers will not certify and allow their mark to be used if the standard is not acceptable. What use is the mark if it is meaningless to any of the players, that is, consumers, industry, and the government.

Some third parties have certified and accredited services and products without first establishing the "buy-in" by the government sector. This has proven to be foolhardy and has diminished the credibility in such certification programs and in the third parties themselves. Likewise, the government has certified services using requirements that were considered technically unreasonable or questionable by industry.

These programs succeed only because they are regulatory requirements. These programs would not succeed if they were advisory and
prescribed only voluntary compliance by industry.

Any question about whether it is legal or beneficial for the government to use a third-party to develop standards can be answered by an
Office of Management and Budget (OMB) circular.[18] This publication,
OMB Circular No. A-119 issued on November 1, 1982 sets forth the
ground rules for federal agency involvement in third-party standards
development. It establishes a policy of preference in this regard. The
following statements are taken directly from the Circular.

> Many Governmental functions involve products or services that must
> meet reliable standards. Many such standards, appropriate or adaptable for
> the Government's purposes, are available from private voluntary standards
> bodies. Government participation in the standards-related activities of
> these voluntary bodies provides incentives and opportunities to establish
> standards that serve the national needs, and the adoption of voluntary
> standards, whenever practicable and appropriate, eliminates the cost to the
> Government of developing its own standards. Adoption of such standards
> also furthers the policy of reliance upon the private sector to supply govern-
> ment needs for goods and services, as enunciated in OMB Circular No. A-
> 76.
>
> OMB Circular No. A-119

Alert and questioning readers are probably wondering if a standard
developed with only industry paying the development cost could be
acceptable to regulators. Please remember the "C" for consensus. The
regulator will not agree to a standard that he finds unacceptable. The
answer is, therefore, that the standard development cost might be paid
for by industry, but the regulator has an equal vote in the adoption of any
standard. Also, it is the regulators choice to use or not to use the
standard and to implement and administer a certification program for its
implementation.

Another question that the reader could ask is "Why does industry
accept third-party certification?" This question is easy to answer because the industries that produce quality products want the others in
their respective industries that produce inferior products to be differentiated from quality producers. Consumers will not buy products that do
not perform well. However, consumers might not be able to inspect
each product for all the attributes that add to the safety or performance
of each individual product, and therefore would not be able to ade-

quately differentiate between quality products and inferior products at the time of purchase.

Industry's answer to the dilemma is to employ the use of third-party inspection and certification. In this manner, good manufacturers can indicate to the consumers that their products comply with standard requirements because they bear the third-party mark of conformance. This protects the industry from competition that produces shoddy goods at low cost and also protects consumers from purchasing shoddy goods.

Why doesn't industry police itself by using trade associations and other vehicles to inspect the members of its own industry? The government has a natural — and well grounded — suspicion of self-regulation by industry. Because self-regulation has little merit in the eyes of the regulators, the third-party is the solution for an industry that is sincere about improving the industry as a whole and supplying good products to the consumer. Another drawback of self-regulation is that it has major complications because of the confidentiality of business information in the industry among competitors. The third-party approach is free of this complication.

Listing Programs

The public has real confidence in some listing programs, notably the NSF and UL programs, because the products that are listed have a long track record of being quality products. The public has confidence in some other listing programs such as those that produce lists of inspected hotels and restaurants. This type of listing is easily understood by the consumer, because most of the companies carefully explain the criteria that the establishments are rated against.

DESCRIPTION OF CURRENT "CERTIFICATION" PROGRAMS

There are several types of "certification" programs in existence. The product certification programs were described in detail earlier in this chapter. These programs entail actual testing of production processes and products for compliance with standards. Other programs exist that are labeled "certification" that are not product certification programs. Three such uses of the term are "certification of individuals", "certification of laboratories," and "supplier certification." A brief description of these systems and others is provided.

Certification of Individuals

Perhaps the most widely known programs for "certification of individuals" are administered by the American Society for Quality Control (ASQC).[19] These programs are for "certified quality engineers (CQE)" and other "certified" quality professional categories. The definition of certification used by ASQC is "the independently verified prescribed level of knowledge as defined, through a combination of experience, education, and examination." They expand this definition further to state that:

> The Certified Quality Engineer is a professional who can understand and apply the principles of product and service quality evaluation and control. This body of knowledge and applied technologies include, but are not limited to, development and operation of quality control systems; application and analysis of testing and inspection procedures; the ability to apply metrology and statistical methods to diagnose and correct improper quality control practices which assure product and service conformity to prescribed standards; an understanding of human factors and motivation; facility with quality cost concepts and techniques; the knowledge and ability to develop and administer management information systems and to audit quality systems for deficiency identification and correction.

The requirements for CQE are proof of professionalism that can be evidenced in many ways, such as registration as a Professional Engineer or membership in ASQC, experience and/or education requirements, and successful performance on a written examination. By definition and description of requirements for CQE, this program assesses the capabilities of a person to perform as a quality services professional.

Certification of Laboratories

The U.S. EPA uses the term "certification"[20] to describe the official EPA approval program of laboratories regulated under the Safe Drinking Water Act and its amendments. The EPA states that this program is based on exact Agency criteria. These criteria, or otherwise called certification requirements, include critical elements on sampling, sample preservation, analytical methods, laboratory facilities, personnel, instrumentation, quality control, data reporting, successful participation in evaluation studies for each analyte to be certified, and an on-site inspection of the laboratory operations based on these stated criteria. The periodic checks of laboratories are conducted only once every three years. The proficiency samples are required to be successfully analyzed once per year. This program is designed to provide evidence of capability, but is not designed to monitor ongoing laboratory performance.

Certification of Suppliers

A new term was recently coined by the Pharmaceutical Manufacturers Association.[21,22] The term, "certified supplier," means that a supplier, after extensive investigation, is found to supply material of such quality that it is not necessary to perform routine testing on each lot received. The concept of "certified supplier" means that a supplier of starting products must pass very specific and exacting criteria for production of materials. These nine criteria, developed by the ASQC's Customer-Supplier Technical Committee are directly quoted:

1. Having virtually no product-related lot rejections for a significant time period.
2. Having no non-product related rejections for a stated period of time.
3. Having no production-related negative incidents for a stated period of time.
4. Having successfully passed a recent on-site quality system evaluation.
5. Having agreed-upon specifications.
6. Having a fully documented process and quality system.
7. Having the ability to furnish timely copies of certificates of analysis, inspection data, and test results.
8. Certification of bulk suppliers requires the correlation and validation of laboratory results so that the customer can utilize the supplier's results as it would its own.
9. The certification requirement for piece part or assembly suppliers is a demonstration of statistical process control.

DESCRIPTION OF CURRENT "LISTING" PROGRAMS

The most well known listing system is the NSF listing program that has been discussed in relation to the NSF product certification program. The two are similar in all aspects. The only difference is that NSF-listed products are assessed for conformance with product standards that are developed by NSF and NSF-certified means that the products are assessed for conformance with product standards that are not developed by NSF.

CRITIQUE — DO THE CURRENT PROGRAMS MEET PRODUCT CERTIFICATION REQUIREMENTS?

The three types of "certification" programs that were presented in the previous section of this chapter are very different from each other. All but one of the programs are vastly different from product certification. "Certification of individuals" and "certification of laboratories" provide only evidence of capability to perform. On the other hand, the "certification of suppliers" is a rigorous program that not only inspects capability of the process to perform, but also carefully evaluates the product rejections and analyzes the types of product failures in addition to non-product failures.

Clearly, only "supplier certification" can be considered to be as rigorous as third-party product certification. This type of program differs from true product certification because it does not involve a third-party. The customer and supplier can work hand-in-hand to implement and monitor the program. This is an area, that would lend itself well to a third-party monitoring the program after it is fully implemented by the suppliers and customers. The third party could conduct inspections and perform routine inspection of conformance of products.

PRODUCT CERTIFICATION AND DATA INSPECTION — THE CRITICAL LINK

Inspection of analytical data is analogous to inspection of a finished product from a factory. In this special case, the laboratory process can be compared with a factory process. The product is the physical evidence of the electronic signals, calculations, and other information recorded about a sample as it is processed in the laboratory. Through careful assessment of the data, it is possible to determine if the "data product" conforms to specifications. Only in this way can one have tangible evidence of the presence of defects and the magnitude of the defects in the product that was produced by a laboratory process. If the purchaser of data knows what defects are present in the data, it is sometimes possible to use it for some purposes but not for others. The product inspection provides information about defects and limitations in the data to the purchaser, so the purchaser might be able to use the "product data" in spite of some flaws. In addition, the identification of the defects sometimes enables the purchaser to understand that the sample or "raw material" cannot be processed by the same analysis method as it was originally processed or the result will again be defective data. The data purchaser then can decide to use the data "as is" for the decision, or to analyze the sample, or a similar sample with a different method to obtain data that does not contain the same flaws.

Inspection of the "product data" to concise specifications allows information to be provided to the customer. A customer that purchases data as a list of numbers as the product obtains little information unless the data are perfect. The likelihood of perfection in many analytical methodologies and with most difficult samples is not high. For this reason, the user of this data is taking a risk in using data that is represented as a table of numbers from a laboratory. Defects and flaws in the data most likely do exist, but the user does not know where the flaws are and how they affect the use of the data.

Laboratory data certification and product certification are exactly the same entity. The criteria for inspection and the article inspected are the only differences in the two types of certification. As with all product certifications, an inspection of the process must be accomplished as an integral part of any product certification program. The purpose and importance of this step in verifying that a process is successfully producing end products is the same for product certification and data certification. Therefore product certification is actually a two step process of "factory accreditation" of the process and "product certification" of a product manufactured by the factory.

REFERENCES

1. "Third Party Certification Program," The American National Standards Institute (ANSI) Z-34.1-1987, Section 4.
2. McClelland, Nina I., " The National Sanitation Foundation's (NSF's) Third-Party Programs to Assist States," presented to the Association of State Drinking Water Administrators (ASDWA), Tampa FL (February 1989) pp. 1,10.
3. Felix, Charles W., and George A. Kupfer, "Third Party Certification — A Measure of Protection for the Consumer," *Journal of Environmental Health* 50(6): 347 (1988).
4. Chumas, Sophie J., Ed., *Directory of United States Standardization Activities*, National Bureau of Standards Special Publication 417, U.S. Government Printing Office, pp. 161-163 (November 1975).
5. Felix, Charles W., and George A. Kupfer, "Third Party Certification — A Measure of Protection for the Consumer," *Journal of Environmental Health* 50(6): 347 (1988).
6. "NSF Starts 1990 With Good Press," *NSF News* (March-April 1990) pp. 2,3.
7. Chumas, Sophie J., Ed., *Directory of United States Standardization Activities*, National Bureau of Standards Special Publication 417, U.S. Government Printing Office, pp. 1-2 (November 1975).
8. "BOCA Exec Defends NSF Copyright," *NSF News*, p. 1. (Fall 1989).
9. Felix, Charles W., and George A. Kupfer, "Third Party Certification — A Measure of Protection for the Consumer," *Journal of Environmental Health* 50(6): 347 (1988).
10. Chumas, Sophie J., Ed., *Directory of United States Standardization Activities*, National Bureau of Standards Special Publication 417, U.S. Government Printing Office (November 1975) pp. 29-30.
11. Chumas, Sophie J., Ed., *Directory of United States Standardization Activities*, National Bureau of Standards Special Publication 417, U.S. Government Printing Office, pp. 30,38 (November 1975).
12. Felix, Charles W., and George A. Kupfer, "Third Party Certification — A Measure of Protection for the Consumer," *Journal of Environmental Health* 50(6): 348-349 (1988).
13. Felix, Charles W., and George A. Kupfer, "Third Party Certification — A Measure of Protection for the Consumer," *Journal of Environmental Health* 50(6): 349 (1988).
14. Felix, Charles W., and George A. Kupfer, "Third Party Certification — A Measure of Protection for the Consumer," *Journal of Environmental Health* 50(6): 349 (1988).
15. McClelland, Nina I., "The National Sanitation Foundation's (NSF's) Third-Party Programs to Assist States," presented to the Association of State Drinking Water Administrators (ASDWA), Tampa FL (February 1989) pp. 4,5.
16. Felix, Charles W., and George A. Kupfer, "Third Party Certification — A Measure of Protection for the Consumer," *Journal of Environmental Health* 50(6): 349 (1988).

17. McClelland, Nina I., "The National Sanitation Foundation's (NSF's) Third-Party Programs to Assist States," presented to the Association of State Drinking Water Administrators (ASDWA), Tampa FL (February 1989) p.2.

18. U.S. Office of Management and Budget (OMB) Circular No. A-119 (November 1982).

19. "Certification-Mechanical Inspector," (Milwaukee, WI:American Society for Quality Control, 1985) pp. 1-4.

20. "Availability, Adequacy, and Comparability of Testing Procedures for the Analysis of Pollutants Established Under Section 304(h) of the Federal Water Pollution Control Act, Report to the Committee on Public Works and Transportation of the House of Representatives, EPA/600/9-87/030 (1988).

21. Maass, Richard, A., "Supplier Certification — A Positive Response to Just-In-Time," *Quality Progress*, (September 1988) pp 75–76.

22. "Vendor Certification: Developing Partnerships Between the Pharmaceutical Industry and Its Suppliers," *Pharmaceutical Processing* 7(2): 20 (1990).

Chapter 5

THE ENVIRONMENTAL PROTECTION AGENCY CONTRACT LABORATORY PROGRAM — A DE FACTO CERTIFICATION PROGRAM THAT INCLUDES PRINCIPLES OF EACH SYSTEM

INTRODUCTION

It has often been stated in a variety of ways that those who do not learn from the past are doomed to repeat mistakes. The Contract Laboratory Program (CLP) is an outgrowth of efforts by the Environmental Protection Agency (EPA) to remedy the wide array of mistakes made in the early environmental sample analysis programs. Instances of these mistakes are legion and need not be delineated here. Suffice it to say that many talented and highly motivated individuals recognized the critical deficiencies of the analytical programs that existed then. The efforts of these individuals culminated in a highly structured contract laboratory analysis program with defined performance criteria. In order to maintain the integrity of this program a continuous quality assurance oversight function was instituted. In this chapter we shall discuss the development of the CLP.

The CLP was designed and developed to support the EPA's Superfund program by providing analytical data of "a known and documented quality."[1] All Superfund related activities were originally authorized under the 1980 Comprehensive Environmental Response, Compensation and Liability Act (CERCLA) and currently under the

1986 Superfund Amendments and Reauthorization Act (SARA). The major objective of the CLP is to provide access for Superfund program efforts, to a range of analytical services on a high volume cost effective basis. Moreover, the CLP, through its highly structured, inflexible protocol procedures, provides legally admissable[2] analytical data for use in EPA enforcement and litigation efforts. The CLP provides access to analytical support by direct response to request from the ten EPA Regions, the major clients of these laboratory services. However, the users of the CLP are expanding to include states, other EPA program offices,and other Federal Agencies.

The overall direction of the CLP is the responsibility of the Analytical Operations Branch (AOB) of the Hazardous Site Evaluation Division (HSED) of the Office of Solid Waste and Emergency Response (OSWER), located at EPA's Headquarters in Washington, D.C. The management function includes the overall coordination of the CLP regarding Superfund objectives; interfacing with the primary user groups; policy and budget formulation and implementation; contract development and administration for CLP services; analytical protocol development and review; review of the analytical sub-contracts and CLP generated analytical data; monitoring and contract performance evaluation; and direction of the CLP QA, in coordination with overall EPA Superfund activities.

For the purpose of our current discussion, the primary focal point of direction within this management scheme is the CLP QA Coordinator. The CLP QA Coordinator manages the many facets of the application of QA procedures to the overall program. As an integral part of this effort, the CLP QA Coordinator works in close concert with EPA's Office of Research and Development (ORD). The administration and research for improving the QA program and procedures is under the purview of the Environmental Monitoring Systems Laboratory in Las Vegas (EMSL-LV). The CLP QA Coordinator works closely with EMSL-LV personnel to ensure improvements in QA procedures are developed, refined and implemented in a timely manner. The CLP QA Coordinator also interfaces with CLP data users to resolve issues of data quality and useability.

Recognizing the broad scope of Superfund activities, a reasonable question arises. This question is, "What procedures are in place to ensure that CLP generated analytical data are useable as is often claimed?" A review of the procedures currently in place would seem to be the appropriate starting point. The key elements of the CLP oversight program can be most effectively divided into pre-contract award and post-contract award elements as follows[3]:

Pre-Award Elements
- Initial Performance Evaluation Sample
- Technical On-Site Evaluations
  - Facilities and Equipment
  - Personnel Qualifications
- Standard Operating Procedures (SOPs) and QA Plans
- Evidentiary On-Site Evaluations

Post-Award Elements
- Periodic On-Site Technical and Evidentiary Evaluations
- Periodic Performance Evaluation Testing
- Data Package Contractual and Evidentiary Assessments by Government Contractors
- Data Package Technical Assessments by EPA Regions
- Data Package and Raw Data Audits by Government Contractors

Each of these elements are now examined in turn to determine how they fit into a coordinated whole.

PRE-AWARD ELEMENTS

Initial Performance Sample

The first criterion for inclusion in the CLP is the successful analysis of a preaward performance evaluation (PE) sample. Any laboratory that wants to become part of the CLP can request a preaward PE sample. A monetary deposit is required before the laboratory receives the sample. This deposit is returned upon submission of the resultant data from the analysis of the PE sample.

These preaward samples, both organic and inorganic, are distributed by EMSL-LV and are intended to be representative of the types of samples routinely analyzed if the laboratory is awarded a contract. The laboratory is required to analyze the PE samples according to the appropriate contractual requirements and report the results of the PE sample analysis. These results are evaluated by EPA personnel for compliance with contract requirements and accuracy of analytical results based on analyte levels in the PE sample. The minimum acceptable score, which is defined by the Agency, provides the basis for the initial segregation of acceptable and non-acceptable laboratory performance. Failure to perform satisfactorily precludes the laboratory from consideration for contract award.

Technical On-Site Evaluation

Following successful analysis of the preaward PE sample and assuming the submitted price per sample is within the EPA-defined competitive range, an on-site laboratory evaluation is conducted. Agency personnel and their support contractors conduct these laboratory evaluations.

The technical on-site evaluations are designed to ensure that potential laboratory participants have the minimum required facilities and equipment and that current laboratory personnel meet the qualifications as outlined in the contract requirements. Furthermore, at this juncture, the laboratory's SOPs and QA Plan are reviewed for consistency and adequacy.

While each of these criteria is critical to the successful generation of useable analytical data, perhaps the elements that provide the best indication of the quality of the laboratory operation are the SOPs and the QA Plan. As such, these elements are examined in further detail.

The CLP is a non-research laboratory program. Therefore, the methods and operations required by the program require the repetitive completion of tasks. Therefore these methods and operations can be effectively governed by a systematic set of operating procedures, usually known as Standard Operating Procedures (SOPs). The quality of a laboratory's SOPs is a critical indicator of the laboratory's ability to function within the CLP. The SOPs that are reviewed by the on-site evaluation team are a reflection of the analytical activities currently performed in the laboratory.

These SOPs must:

- be consistent with EPA regulations, guidelines, and contractual requirements
- provide evidence documenting the performance of all tasks required to successfully complete a sample analysis according to the appropriate protocol

- provide a system to demonstrate the validity of analytical data generated by the laboratory and to explain the cause of missing or inconsistent data
- describe the procedures to implement corrective actions and the feedback procedures used when analytical data do not meet minimum requirements
- be updated as required by changes in procedures or required schedule, as appropriate
- be stored for future reference
- be available to all personnel involved in the generation of analytical data
- be part of a document control procedure sufficient to prevent the use of outdated or inappropriate SOPs

The EPA requires SOPs for sample receipt, storage, and preparation, glassware cleaning, instrument calibration and maintenance, analytical standards receipt, storage and use, and data reduction, validation and documentation procedures. The quality and completeness of these SOPs are an extremely important indicator of the laboratory's ability to perform the required analytical work and the laboratory's understanding of its function within the CLP.

Laboratories that participate in the CLP must establish a QA program with the overall objective of providing useable analytical data. This QA program must minimally incorporate all quality control(QC) procedures, corrective action procedures, documentation involved in data collection, and all QA procedures performed to ensure the generation of useable data. The laboratory must prepare a fully documented QA plan. This QA plan documents all procedures designed to achieve at a minimum:

- documentation of all aspects of the analytical data generation process
- detection of problems associated with analytical data generation and corrective action procedures designed to maintain reliable analytical data
- procedures in effect to maintain a qualified laboratory staff
- procedures to ensure that all analytical instruments and systems are maintained in an appropriate state of operation
- procedures to maintain data integrity, validity, and useability
- procedures for self-inspection of all appropriate laboratory operations

It is imperative that the quality assurance plan documents the laboratory organization, policies, objectives, guidelines and all appropriate QA and QC procedures used to ensure data quality and useability. This QA plan must contain the organization of the laboratory personnel and their qualifications, facilities, equipment, sample and document control, analytical methods, data generation and review procedures and the self-inspection system.

Due to the critical nature of the QA activities associated with the generation of acceptable data, a thorough review of the laboratory's QA plan will provide strong evidence of the laboratory's ability and dedication to the generation of quality analytical data. Any organization that perceives QA activities as a necessary evil and an overhead function that must be endured will not function well in the CLP. Successful organizations correctly view QA activities as a key function of an integrated quality approach.

Evidentiary On-Site Evaluation

The National Enforcement Investigations Center (NEIC) performs a preaward evidentiary on-site evaluation to determine if laboratory policies and procedures are in place to ensure that all evidence handling requirements are met.[4] Due to the potential for all Superfund analytical data to become a pivotal factor in litigation activities, it is vital that evidentiary aspects of laboratory operation be adequately addressed. As with SOPs and the QA plan, the understanding of the laboratory regarding the importance of evidentiary chain-of-custody and documentation requirements is a critical indicator of the laboratory's potential for success in the CLP.

Separate evidence requirements and audits are not usually a part of other accreditation and certification programs. It is imperative to stress the importance of this type of inspection so that the requirements are incorporated into a credible national system. Samples are physical evidence that are collected. It is vitally important that samples and their results are correctly matched and that there is no way that the samples can be altered in the process from collection to final data reporting. It is also important to have a record of who handled the samples and that they could not have been tampered with during the time that they were taken and the time that they were analyzed. In addition, the data that is produced for the samples must be complete, and must be sufficient to allow reconstruction and reinterpretation of that data, if necessary, by technical experts. If the data cannot be shown to be obtained from analyses of unaltered samples, or if the necessary documentation to appropriately support the data cannot be provided, there could be problems in using the data in a court of law.

POST-AWARD ELEMENTS

Periodic On-Site Technical and Evidentiary Evaluations

Following award of a CLP contract, the laboratory becomes an integral part of a continuous QA oversight program for the life of the contract award period. At a frequency determined by the Agency, EPA personnel and their support contractors conduct on-site laboratory evaluations. The frequency of on-sites is determined by a quarterly analysis of trend data that includes the numbers of samples that the laboratory analyzed in the past, scores on PE samples, evaluations of data packages, compliance with the contract, and audits of raw data. Evidentiary audits are scheduled based on other criteria, such as changes in location of a laboratory, data package defects, changes in personnel, and time since the last audit. On-site evaluations consist of two parts — technical and evidentiary. The two parts may not be scheduled at the same time, because of the different criteria on which the scheduling is based.

Technical evaluations consist of inspections that are similar to the preaward on-site visits that were described previously in this chapter. However, post-award on-sites can be more effective because the trend information from actual performance of the laboratory on samples allows assessors to focus on expected laboratory difficulties. These on-site visits also allow the assessors to determine if the laboratories' self-inspection systems are working properly. The assessors have knowledge of the systems' effectiveness before the on-site visits, but the visits can clarify what the problems are. The on-site reports of deficiencies can provide guidance for the laboratories on the actual points of departure from requirements. The on-site visits are also useful to verify that laboratory personnel have not changed and that the laboratory facilities and instrumentation are maintained as required in the contract.

At the end of the on-site visits, the laboratories are informed of the findings of the evaluation team by the appropriate contract monitor. The official findings are reported in technical on-site evaluation reports to the CLP contract Project Officer and Deputy Project Officer. These reports delineate the contractual deficiencies that were observed at the on-site visits. The laboratories will be instructed by the appropriate contract Project Officer or Deputy Project Officer that corrective actions must be taken, if necessary.

Evidentiary audits are designed to determine if the laboratories' policies and procedures are in place to address evidence handling requirements. Evidentiary audits are comprised of the following: a procedural audit consisting of a review and examination of actual SOPs and documentation, a review of SOPs to determine the accuracy and completeness of the SOPs for sample receiving, sample storage, sample tracking, and analytical project file organization and assembly; and an analytical project file evidence audit consisting of a review and examination of the analytical project file documentation.

Upon completion of on-site laboratory evaluations, technical reports and evidence audit reports are prepared that delineate any deficiencies found. These reports are provided to the appropriate EPA personnel and the laboratories. The laboratories are expected to address deficiencies and to provide a course of corrective action.

Periodic Performance Evaluation Testing

Perhaps the single most critical element of the on-going QA oversight program is the requirement that the CLP laboratories periodically analyze PE samples. This is the one truly independent means of measuring the performance of both the laboratories and the analytical protocols used by the CLP. Laboratories participating in the CLP must analyze PE samples on at least a quarterly frequency. These PE samples could be either single or double blind. The program strives to send samples that are representative of samples analyzed routinely, but many times these samples consist of spiked water samples because of budget and technical considerations.

The results of the analysis of the PE samples are evaluated, at a minimum for compound identification, quantitation, and sample contamination. Quantitation of target analytes is defined based on reported values using population statistics. Laboratories are required to use the NEST Mass Spectral Library to tentatively identify a number of non-target compounds that might be present above a minimal response. Should the laboratories fail to perform satisfactorily when analyzing these PE samples, the laboratories might not receive CLP samples for analysis pending analysis of remedial PE samples or other corrective actions. Obviously, it is in the best interest of the laboratories to perform the PE analyses in a manner consistent with the generation of high quality analytical data.

Package Contractual and Evidentiary Assessments by EPA Headquarters

Following completion of analytical activities and report generation, the laboratories forward copies of the completed data packages to EMSL-LV, the Regional clients, and the Sample Management Office (SMO). SMO is an EPA contractor performing a variety of support functions including sample tracking and invoicing and other financial management activities. One of the more important of these functions is conducting contract compliance screening (CCS). This activity is one aspect of the Agency's contractual right of inspection of analytical data. This process is designed to examine and determine the laboratory's adherence to contractual requirements. Contractual requirements reviewed for compliance include: sample holding times, gas chromatography/mass spectrometry (GC/MS) tunes, initial and continuing calibrations, blanks, and surrogate compound recoveries.

In addition to the assessments of data for contractual and technical compliance, the data might also be inspected for evidence purposes by NEIC or regional personnel. Each case of data must be assembled by the laboratory so that all relevant documents can easily be inventoried by the government. Documents that must be assembled and forwarded to the government include logbook pages, bench sheets, mass spectra, chromatograms, custody records, library search results and all other relevant records. These records are needed to provide document accountability of the completed analyses.

Data Package Technical Assessments by EPA Regions

The majority of all CLP data are generated to meet the need of the ten EPA Regions. Consequently, these Regional clients are extremely interested in the quality of the analytical data. Because the Regions have the final decision regarding the useability of the data for its intended purpose, each Region has developed a procedure for review of analytical data. These review procedures have been developed using the general guidelines agreed upon by the Regions and AOB.[5,6] As the basis for data evaluation, the perspective of end users is of paramount importance in the review process. Each Region can enhance the basic guideline approach with additional review based on region- or site-specific factors.

As data users develop specific requirements for the data that they use for decisions, they can develop specifications for the generation and review of data so that it is more relevant to their particular needs. Superfund is promoting this user-interface in specifying data needs to the CLP and data reviewers by promoting "Data Useability" workgroups and definition of data requirements by data users.[7] This is a positive step to promote accountability for appropriate data acquisition. Laboratories cannot produce useable data unless decision makers request the correct type of data and specify critical requirements of that data.

Data Package and Raw Data Audits by EPA EMSL-LV

On some infrequent basis, EMSL-LV personnel and their support contractors conduct comprehensive reviews of a subset of CLP sample data packages using a Military Standard 105D[8] approach. In contrast to the CCS approach, this review is of a technical nature and is designed to identify problems occurring within the laboratories or with the analytical protocols. Reports arising from these technical reviews are provided to the laboratories and the appropriate EPA personnel. These reports serve a dual purpose — problem identification and as a point of discussion — for the laboratories and EPA to resolve the problems identified.

Perhaps the single most inclusive review of the raw analytical data is accomplished by the GC/MS magnetic tape audit. Support contractors for EMSL-LV conduct these audits. Depending on the nature of the tape audit, any or all of the raw data and quantitation reports for samples, blanks, surrogates, matrix spikes, matrix spike duplicates, initial calibrations, continuing calibrations, and instrument tunes associated with the audited sample cases can be scrutinized. In order to reference the raw data on the GC/MS tape with the appropriate reports, user generated spectral libraries, extraction laboratory bench sheets, instrumental references, and logbook pages are also reviewed. A thorough review of the raw data provides a wealth of information — not only regarding the performance of the laboratories, but also about the utility, precision, and accuracy of the analytical protocols.

CLP STRENGTHS AND WEAKNESSES

The CLP is, without a doubt, the single most comprehensive analytical program yet devised. As previously described, it is a program designed to address all aspects of the generation of analytical data. The comprehensive requirements are the basis for its great strength and simultaneously contribute to its weaknesses.

Perhaps the greatest single positive attribute of the CLP is the continual oversight of the participating laboratories during the period the laboratory is producing analyses under a CLP contract. This oversight is multifaceted, as previously described, and is designed to review every component of the data generation system. Consequently, this system has the inherent capability of providing information that is useful for solving problems encountered in the analysis of environmental samples.

Theoretically, problem solving can be most effectively accomplished by providing a feedback loop between the group responsible for the QA oversight and the participating laboratories. The information to drive this problem solving mechanism arises from results of the periodic analysis of PE samples, review of data packages (both for technical compliance and for contract compliance) data useability reviews, and so forth. By informing the participating CLP laboratories of their performance in these critical areas, and most importantly by assisting the laboratory in correcting the deficiencies identified during this QA oversight, it is possible to ensure that analytical data are of an acceptable quality. Obviously this presupposes that both the EPA and the participating laboratories are mutually interested in solving the problems identified.

In the CLP, the results of information about causes of problems in laboratory performance are not directly conveyed to the laboratory. In fact, the CLP cannot suggest improvements that might be obvious to the on-site assessors for contractual reasons. Therefore, the information that could assist laboratories in correcting problems and generating better data, is many times not available to the laboratories as constructive criticism. Instead, information about problems is conveyed in findings reports, which point out the defective conditions, but offer no assistance in determining corrective actions and making improvements.

A key strength of the CLP is that all data are reported in a concise format. There is wide disagreement on what elements a data package should contain. However, in the absence of any agreement on minimum data deliverable requirements, the CLP requires specific data deliverable requirements. This standard format for all data enables a structured review of deliverables. Lack of consistency in data deliverables gives rise to questions about the meaning of detection limit, quantitation limit, blank data, sensitivity of the analysis, working range of the method, and all other questions that constantly arise when trying to interpret data. Data that is supplied in other formats often is unusable because such questions cannot be resolved.

Consistent format also allows computerized assessment of standard data requirements and facilitates use of computerized expert system approaches for determining useability of data based on the extent and nature of the flaws in the data.[9] Computers can assist only on data that is consistently reported in a well defined format. If the industry does not agree with the standard CLP data deliverable format, it is suggested that a standard format designed to be concise and contain sufficient information to determine useability of data be defined by industry. In the absence of such a format, data cannot be interpreted and used effectively.

The CLP has gathered vast quantities of method-specific, analyte-specific, matrix-specific, laboratory-specific, and other information that can be used to improve laboratory methods and data. In addition, these data can be used to define method-research needs and the most cost-effective quality control requirements. These data have not been fully utilized for these purposes. However, the strength in the program is that the data were collected and can be used in the future to strengthen environmental data programs and use.

Another strength of the CLP is sometimes perceived as a weakness. The CLP requirements are specific, detailed, and sometimes considered onerous by the participating laboratories. However, it is because these requirements are so specific and documented that evidence of non-compliance with the requirements can be determined. This idea is addressed more fully in Chapter 14, but it is mentioned here because omission would be perceived by many as an oversight of a weakness of the CLP. The Office of the Inspector General (OIG) is investigating if some CLP laboratories might not have been compliant with the terms of the CLP contracts in the past.[10] Because of this, many laboratories have been, or are currently under, investigation by the OIG. In some cases, the OIG might determine that the laboratories have committed serious wrongdoing under the terms of their contracts. This same type of OIG investigation has not occurred in other certification and accreditation programs. Therefore, critics of the CLP use the OIG investigations to indicate that the CLP is a failure. This does not prove that the CLP is a failure. In contrast, the fact that wrongdoing can be found, documented, and resolved states that the CLP is a success. It is the only program for which the requirements and contract documentation are well enough established that discovery of contractual waste, fraud, and abuse by the OIG is possible.

The weakness that is suggested by the OIG investigations that are discussed in the previous paragraph do indicate one weakness of the CLP. Unfortunately, not all CLP related activities have resulted in success. Since its inception, the CLP has evolved from a program designed to provide high quality analytical data into a contractually driven program. Inflexibility, partly based on contractual difficulties and opportunistic adherence to contract requirements by the laboratories, have sometimes resulted in the generation of questionable analytical data. Some requirements in the CLP are perceived by laboratories to have little technical merit and are sometimes disregarded on that basis. The low bid prices for EPA samples sometimes are used to justify "cutting corners" in cases where the laboratory determines that the CLP requirements are technically not important to the quality of the data. However, since the CLP is a contractual program, the laboratory must not disregard any contractual requirements. The OIG investigations do not differentiate between infractions that severely impact the technical validity of the data and those that do not. This is because the object of the investigations is to determine if laboratories have disregarded any contractual requirements.

Some critics of the CLP complain that the system produces data of "known and documented" quality, but that these data are not suitable for use in supporting many types of Superfund decisions. This is not necessarily a weakness in the CLP QA system, but is a weakness in the CLP. The users of data must define their needs and communicate these needs to the AOB so that the CLP obtains the correct types of data. It could also be a weakness in AOB's ability to retain the current contract structure and still meet the data needs of all data users. The CLP was designed to produce one type of data. As data users become more sophisticated and can define their data needs better, the CLP will be asked to produce many different types of data. The ability to support many different data user's needs within the current contract constraints is difficult. Yet, this is not a weakness in the CLP QA system as it was designed. It is a weakness that must be addressed for Superfund to obtain all necessary types of data in the future. The AOB support of the data useability workgroup[11] effort is evidence that they acknowledge the problem and are attempting to address it.

One weakness of the CLP exists because the program is highly structured and very large. The associated analytical protocols have become rigidly structured with only minimal opportunities for rapid incorporation of improvements reflecting changing technology. In this regard, the CLP is typical of many regulatory systems. On one hand, the structure of the CLP allows for easy assessment of data and determination of when and how contract deviations occur. This structure also allows interpretation of the data to determine how the defects in the data affect possible use of the data. On the negative side, highly structured inflexible approaches are slow to change and are often not state-of-the art if they are driven by contractual requirements. Contracts cannot quickly be changed to allow the best science to be accomplished.[12] The use of highly structured contracts fosters a prevalent "status quo" attitude and can generate an adversarial atmosphere in which cooperation between the EPA and its contract laboratories is difficult. Too often, good ideas are forgotten, or not included in new contracts because of the length of time that passes between the suggestion of relevant changes to the contracts and the time that the contracts are rewritten and reissued in the bid process. This weakness must be acknowledged when designing a national accreditation and certification system.

One final weakness, and perhaps the most important weakness, is that the CLP is limited to only two specific methodologies. The program is limited to acquisition and oversight of gas chromatograph-mass spectrometer (GC-MS) and GC organic methodology, and inductively coupled plasma arc (ICP), ICP-MS, and atomic absorption (AA) inorganic methodology. (Dioxin analysis is also included as a routinely available service, but the data deliverable inspection is different from the other two services, so is not included in this discussion.) The CLP system was designed to include only oversight of contract-generated data. As we attempt to generalize the CLP approach to all environmental data acquisition, the problem becomes apparent. How can oversight of all environmental data (non-CLP) data be provided? The approach to inspection of non-CLP data can be designed only if requirements for all Superfund data are specified. In addition, an assessment of the capability of laboratories to perform Superfund analyses cannot occur until minimum laboratory requirements are delineated. To date, these requirements have not been issued as policy. It is obvious that these steps to define basic requirements will be a challenge for a national accreditation and certification program. The implementation of a CLP-like approach might be difficult to implement, given the well-defined and limited scope of the CLP oversight program as contrasted with oversight of all analytical data.

After listing the shortcomings of the CLP, it is only fair to state that regardless of its failures and obvious shortcomings, the oversight program of the CLP is the best in the industry. Not only does it provide continuous oversight, but it has the inherent ability to provide information critical to solving its problems. It also has all the attributes necessary to become a leader in all areas of analytical data generation and information utilization. This is readily attested by the fact that many corporate purchasers of environmental data require a laboratory to have a CLP contract (be "de facto CLP certified") before they will use the laboratory for analytical support. Especially in the absence of a national certification program that is considered credible by industrial purchasers of data and the government, the de facto CLP certification is the only accreditation that is readily accepted for environmental laboratory data. However, the recognition of the CLP as a de facto accreditation program coupled with the lack of a credible national program causes the environmental laboratory program many problems.[13]

Many of the problems with the CLP that are recognized are not a fault of the QA system. Many exist because of the contractual nature of the program, the lack of integration of the sampling process with the analytical process, and the lack of attention in the Superfund system with identifying analytical requirements of data based on the uses of the resultant data. The QA process used by the CLP should be considered apart from the perceived weaknesses in the entire CLP system. It should be considered apart from the entire system to allow for fine-tuning based solely on possible flaws that it might have. In any case, it is possible to sort through the perceived weaknesses of the CLP and identify causes of the weaknesses only because the CLP QA system is working. If some assignable causes for lack of quality are not currently being identified by the CLP QA system, the system can be improved to allow for identification of the causes. If the CLP QA system contains costly redundancies, the redundancies should be removed before designing another system that is based on this program.

SUCCESSFUL APPLICATION OF THE CLP

The CLP is not the only example of successful use of these procedures. One other example of the successful implementation of this approach is the Love Canal Habitability Study. This study did not suffer from the CLP's inability to relay performance information to the laboratories. In the Love Canal Study, the assessment of laboratory data was immediately relayed to the participating laboratories. While the analyses were not performed by using the normal CLP routine analytical services, the QA oversight was modeled after the CLP QA program. It was imperative that the data generated during this study be of the highest quality. This was accomplished by a continuous interaction between the group responsible for the QA oversight (EMSL-LV) and the participating laboratories. Deficiencies identified were immediately relayed to the laboratories and corrective action procedures were developed jointly by EPA and laboratory personnel. Upon demonstration of problem resolution, the laboratories were permitted to resume sample analysis. This system was instrumental in the success of this critical study. The end result was analytical data acceptable to all concerned parties. This example provides evidence that the CLP QA principles can be successfully applied to other data collection activities.

It is interesting to note that the Love Canal Habitability Study has been challenged[14,15] recently because the study failed to address the site's health risk. This site is the one that "spawned the Superfund program" and has received much media attention and scrutiny. However, the actual data quality produced under the CLP-like QA program has not yet been questioned.

PUBLIC PERCEPTION AND CONFIDENCE AND SCIENTIFIC VALIDITY

While it may be true that the fates dice with the destinies of men, the fate of our world depends on decisions made by mere mortals. Moreover, these decisions are dependent entirely on the quality and useability of the data and information provided to the decision maker — whomever they may be.

The heightened awareness of the public regarding problems requiring analytical data has resulted in a more participatory role of the public in the decision making process. This is laudable and must be encouraged if technology is to be viewed in a positive vein. Many instances come to mind of the successful integration of technology and the public welfare. The entire National Aeronautics and Space Administration program is a prime example of positive public perception in a technological program. However, the dark side continues to far outweigh these successes. Paramount among the more widely perceived failures is the nuclear power industry, the abhorrent state of hazardous waste disposal and the adverse impact of industrial activity on our environment.

As a direct result of some Superfund activities, and as an indirect result of the CLP, the general public has a greater respect for data and information provided by government agencies. However, there is a widely held perception that data providing entities — be they Federal agencies or corporate in nature — are suspect. This is simply because of the self-interest inherent in stating that data are useable for the intended purposes. Unfortunately, no matter how intensive or well run, no self-inspection system will ever be totally accepted by those who perceive self-interest as a prerogative of the data generators. The CLP, for all its strength and good intention, is not immune to these doubts. This will be a constant problem as long as the EPA (or any other Agency) provides its own oversight.

To overcome this perception of self-interest, it is necessary to provide independent oversight and technical assessment. Many groups have been suggested for this role, though all suffer from inadequacies, not the least of which is the limited acceptance of the results of these groups. In order to overcome this perception, it is imperative that this oversight entity be truly independent, technically competent, acceptable to all concerned parties and accountable for its decisions. Regrettably, no such group currently exists for the oversight of the generation of analytical data. Pending the creation of such an entity, the public will always doubt the sincerity of self-inspection systems and the validity of analytical data. Enlightened self-interest will provide the impetus for all data users and generators to subscribe to an independent oversight process. This type of self-interest must be encouraged for only then will the public have confidence in the data generated by federal, state, and corporate entities.

COST AND BENEFITS

The result of using questionable data in making decisions that could potentially cost hundreds of millions of dollars and affect the lives of millions of people is incalculable. Many instances exist of costly decisions resulting from the inadequacy of analytical data. They are beyond the scope of the present discussion. Assuming that it is a given fact that inadequate data generally results in questionable decisions it is far more important to stress the effectiveness and benefits of the CLP.

It is also important to stress that data that are known to be of a designated quality, even if they might be flawed in some aspect, are much more valuable and useful than data that are of unknown and therefore, suspect, quality. The cost of the CLP QA program is insignificant compared to the cost of making an incorrect decision based on data that is of unknown quality.

EFFECTIVENESS AND INTEGRITY

Effectiveness and integrity are immutably linked since any QA program is dependent on the integrity of the personnel, both technical and management, involved in conducting the program. Without doubt, the majority of EPA personnel want the program to succeed. Just as obviously, not everyone can adequately fulfill their job functions to make the program succeed well. This is unfortunately the result of limiting the number of government personnel associated with the program and the necessity of the increasing reliance on contractors to perform duties previously the purview of government employees. The program can work, given appropriate contractor support, but the government must retain appropriate oversight of this resource. Some aspects of the oversight may be better performed by a third-party system. This would allow the government and its contractors to focus on the aspects that can best be accomplished within the government's program.

A direct result of this situation is the continuing question of both the effectiveness and integrity of the CLP. Questions have arisen concerning the future of the CLP. As long as the CLP is considered a de facto certification program — a situation denied by Agency personnel — many laboratories will strive to become CLP members, solely for the advertising benefit. Moreover, if the EPA continues to provide only self-inspection, then the programs effectiveness and integrity will continue to be questioned.

Ensuring both the effectiveness and unquestioned integrity of the Superfund analytical activities will require the EPA to undergo a positive reassessment of the current program. The question is simply this, Does the Agency maintain the "status quo" and ensure the escalation of current problems or does it function in an anticipatory mode and ensure the success of the Superfund analytical programs? Ultimately, the use of an independent, technically oriented, widely-accepted, problem solving third-party accreditation program that is augmented by Superfund-specific oversight is the only logical way to ensure a successful Superfund program. Such a system could include both CLP and non-CLP analyses, and therefore fully encompass Superfund's needs.

Specifically, will the Agency respond in a positive visionary fashion or succumb to the seduction of turf battles and the "not invented here" syndrome. Time is limited and the price of indecision is failure. Given that the CLP is the most effective QA program for environmental laboratory data, if it is to retain this reputation, it is imperative that the system change to keep pace with new demands or merges with a more national system.

REFERENCES

1. "User's Guide to the Contract Laboratory Program," U.S. Environmental Protection Agency, Office of Emergency and Remedial Response, Washington, D.C. (December 1986).
2. "User's Guide to the Contract Laboratory Program," U.S. Environmental Protection Agency, Office of Emergency and Remedial Response, Washington, D.C., p. 2. (December 1986).
3. "User's Guide to the Contract Laboratory Program," U.S. Environmental Protection Agency, Office of Emergency and Remedial Response, Washington, D.C., pp. 95–104 (December 1986).
4. "User's Guide to the Contract Laboratory Program," U.S. Environmental Protection Agency, Office of Emergency and Remedial Response, Washington, D.C., pp. 81–84 (December 1986)
5. "Laboratory Data Validation Functional Guidelines for Evaluating Organics Analyses," U.S. Environmental Protection Agency, Hazardous Site Evaluation Division (February, 1988).

6. "Laboratory Data Validation Functional Guidelines for Evaluating Inorganics Analyses," U.S. Environmental Protection Agency, Hazardous Site Evaluation Division (July 1988).

7. Longest, Henry, L., Memorandum: "Improved Flexibility and Quality in Non-CLP Laboratory Data — Request for Workgroup Participation," (April 6, 1989).

8. "Sampling Procedures and Tables for Inspection by Attributes," Military Standard 105D (MIL-STD-105D) Chg Not 2 (Philadelphia, PA: The Naval Publications and Forms Center, March 1964).

9. Pandit, Nitin S., John Mateo, and William Coakley, "IQAP: An Intelligent Quality Assurance Planner for Environmental Data — Functional Requirements," presented at the American Chemical Society Expert System Conference (1989).

10. Himelstein, Linda, "Superfund Effort Jeopardized by Suspect Data," *Legal Times* (April 23, 1990).

11. Longest, Henry, L., Memorandum: "Improved Flexibility and Quality in Non-CLP Laboratory Data — Request for Workgroup Participation," (April 6, 1989).

12. "User's Guide to the Contract Laboratory Program," U.S. Environmental Protection Agency, Office of Emergency and Remedial Response, Washington, D.C., p. 82 (December 1986).

13. Fisher, Steven S., "Letter to the Editor," *Environmental Science and Technology*, 23(1): 5–6 (1989).

14. "Reilly, Cuomo Asked to Block Love Canal Resettlement," *Superfund* 4(7): 1,5 (1990).

15. "Love Canal Homes Are Safe," *Chemical and Engineering News*, p. 24 (May 1990).

Chapter 6

THE TOTAL QUALITY
MANAGEMENT APPROACH

DESCRIPTION

What is total quality management? What can it add to our understanding of accreditation and certification? Up until this point, all discussions about accreditation, certification, listing, and so on have centered around verification of process, product, service, or individual for conformance to requirements. All certification and accreditation systems have been instituted to determine if products, services, individuals and other related entities satisfy certain quality requirements. Accreditation and certification programs are formalized procedures that are accomplished for the purpose of making an unbiased determination of whether the goods or services conform to specifications.

How do accreditation and certification fit into the total quality management scheme? Before this question can be answered, the concepts of total quality assurance must be explained. The history of quality assurance and quality control practices that have evolved into total quality management in this country are now briefly examined.

Early in the history of manufacture of items for trade or purchase, each individual craftsman checked his own work to make sure that he preserved his reputation for making a quality product. Later, as small craftsmen banded together to form small production areas, inspectors were employed to inspect all articles for conformance to agreed-upon workmanship practices. Each article was painstakingly inspected to assure the craftsmen that all their collective efforts produced quality products.

With the advent of mass production at the beginning of the industrial revolution, it became apparent to manufacturers that it was too costly to inspect each article for conformance to quality requirements at the end of the process.[1] The practice of "inspecting in" quality at the end of the manufacturing process caused rework or scrapping of the articles found to be defective. The factory that produced these defective articles has now been termed the "hidden factory". Manufacturers realized that the "hidden factory" could produce a large percentage of the articles made in the plant and that all articles that it produced were defective. This realization made manufacturers understand that the process of tail-end inspection was very costly. If a large percentage of all articles manufactured must be put through the "good factory" again or scrapped, it became apparent that the products of the hidden factory, that is, the defective products represented losses, rather than potential profits. It became necessary to improve the actual production process to avoid costly rework of final products so that all products manufactured were good.

Statistical methods for quality control were then implemented to allow workers to stop production before the process goes out of control and produces faulty products. Statistical process control is still used today and enables optimization of processes in today's factories. The use of statistical process control data has enabled engineers to understand more about the causes and cost of failures. Based on statistical process control information, either the processes are changed or the products are redesigned to minimize the types and numbers of problem areas in the manufacture of products. In addition, products are redesigned to minimize the numbers of parts that each contain. Consequently, conformance to specifications to ensure correct fit with other parts is enhanced. The number of different processes are minimized to reduce the numbers of possible problem areas that must be tracked in the production of articles.

Because statistical production control techniques alert workers that the production process limits are out of specification and because manufacturing design practices have been improved, the practice of 100% inspection of each item has become less critical to assure that the quality requirements of an article are met. The use of statistical sampling techniques to determine if a process is producing items according to specification were introduced. These sampling techniques are based on sampling of the entire production lot and then extrapolating the results of the inspection of the sampled articles to the entire lot.

Many quality assurance professionals view past practices with interest and also see that the only way to assure defect-free production is to design processes that are not capable of producing defective materials. This is quite a tall order, but it is what total quality assurance is all about. In order to begin to make products that are free of defects, the entire processes that make products must be examined and reexamined with the object of discovering all possible sources of defects. This is a never ending process and involves all people that are involved in the processes. The quality assurance professional's role in this new age of total quality assurance has thus become a facilitator of excellence rather than an inspector to specifications.[2]

In examining the processes that are necessary for production of goods and services, it is apparent that there are three distinct parts of a quality system. The first step in the quality system is engineering or planning the production process so that it is capable of producing a quality product. Next comes the production of the product by effective manufacturing techniques and by employing statistical process control to the manufacturing process. Finally, comes the inspection process to determine if the product meets the specification requirements. Final inspection should employ appropriate statistical sampling techniques. The total quality system consists of the collective plans, activities and events that are provided to ensure that a product, process or service will satisfy given needs. The quality system includes all planning, design, statistical process control, inspection and sampling techniques employed for quality assurance and quality control purposes in the production of a product or service. The quality system includes all the elements of quality assurance and quality control in an organization. The quality program is the documented plan for implementing the quality system.[3]

In the past, the term total quality management was not specifically used by industry. The term that was often used in the past that comes closest to representing what this new term means is the term, quality program. Quality programs are the written documents that delineate how to implement the elements of quality assurance and quality control practices that allow companies to produce quality products. Quality programs were, in the past, the keys to improvement and measurement of quality.

Quality programs are still the keys to quality improvement and management. However, management's understanding of quality assurance is changing. Now, quality functions are no longer relegated to an inspection department. In the past, much emphasis was placed on evaluating products according to static requirements. Today, the emphasis has changed to improving quality and constantly meeting more stringent quality requirements. This is quite a challenge, and a much bigger job than past practices of compliance with static requirements. This planning of design, planning of production, and planning of inspection is a management function. The process of discovering all waste and error in processes is a management function. The term total quality management is a reflection of this change in status for quality assurance functions.

The ways of measuring quality expenses are improving. As a result of knowing better the actual cost that errors and rework represent, management's attention is becoming more focused on prevention of quality problems, rather than fixing the problems after they have occurred.[4] The management of the entire process of better design, better production, and better and less need for inspection is now called total quality management.

The quality control manager is no longer the "keeper of quality" but is rather a facilitator and teacher that assists each person in understanding what quality is and how to improve quality in their own discipline, specialty, and functional area. This change in thinking about quality from seeing it as a function in the company to a focus of the company is integral to the long term improvement of our companies. The transformation to managing with quality as a focus requires a new set of principles for operating in this new business climate.

UNDERSTANDING TOTAL QUALITY MANAGEMENT — DEMING'S FOURTEEN POINTS

W. Edwards Deming[5] has provided an easy to understand set of principles that are commonly known as Deming's 14 points to management. Deming is well known as a leader in the quality disciplines and is noted for his understandable and forthright style of educating managers on effective means to manage for quality. He also points out clearly in his teachings that managers are responsible for the company quality function. His approach of management for continuous improvement of quality leads to lower cost, improved quality, and higher productivity. Deming and other quality professionals such as Joseph M. Juran[6] and Kaoru Ishikawa[7] agree on many of the basic elements of quality improvement. Deming's approach to total quality management is presented here because it is simple, concise, and is one of the better known total quality approaches. All 14 points are listed to provide the basis for understanding total quality and how accreditation, certification and other inspection systems fit into the new total quality management approach.

The 14 points are expanded here to show their importance and relevance to both laboratory operations and for the accreditation and certification programs that evaluate the quality of laboratories. In some cases, the principle itself is self-explanatory. In such cases, the authors have pointed out the impediments to the implementation of the principle that current practices in the laboratory industry have caused. In addition, for cases in which the industry has implemented the principle and quality improvements are evident, the success story is provided.

It is important to note that the laboratory industry is not the only one in which total quality management has been difficult to implement. Many large, contractual quality assurance programs are also plagued with similar problems.[8,9] These problems occur because the systems were designed primarily as inspection systems that provide assurance that procurements meet requirements. Because the systems focus only on end product inspection, they do not include the principles of statistical process control and total quality management.

DEMING'S 14 POINTS

1. Create Constancy of Purpose Toward Improvement of Product and Service.

The company must create constancy of purpose for the improvement of the company. Only in this way will the company become excellent, produce products that satisfy the customers, and to provide jobs for the people that work to make the company an excellent firm. This requires constant and continuous improvement, long-term planning, and innovation and constant hard work from all persons in the company. To achieve this point, the management must:

- Enlist the trust and support of all persons in the firm.
- Define the goals of the company for all to understand.
- Define the processes that are required to produce a product.
- Define the standards for development, production, and service for the long-term, rather than just the short-term.

- Invest in capital equipment and education of operators to support the goals and production standards set forth by management
- Define the internal and the external customers. Define the needs of these customers. Define what "service to customers" is.
- Develop better ways to provide the products and services in less time, using fewer resources.

There are several reasons that environmental laboratory managers are unable to fully implement point one of Deming's approach. The most crippling reason is because the managers cannot implement continuous process improvement, except under strict constraints. The environmental laboratory business is required to use set procedures that do not allow innovation or improvements because they are considered deviations from required procedures. The required procedures set by the EPA and other regulatory agencies define the exact procedure by which samples must be processed, rather than defining the performance requirements of the product data. Therefore there is little incentive to improve processes and procedures. In fact, the laboratories are penalized for any type of process improvement because it is deemed a deviation from the required procedures. Since there is one required way to make the product, the laboratories cannot change processes to improve profit margin. The laboratories cannot use better equipment or processes to improve the product because of these constraints. The strict procedure requirements make any innovation and improvements to procedures a foolish expense. If the laboratory managers cannot employ improved techniques and procedures in the business, it does not make sense to expend resources to develop the techniques.

Capital equipment purchases are not made on a long-term basis, but are dictated by the changes to required process requirements that occur on a short-term basis. Staffing plans for laboratories that process mainly government contract samples are handled on a short-term basis because the day-to-day workload from these contracts is not stable and because the contracts are awarded on a lowest-cost basis, (with some consideration of capability). In addition, the contracts are not necessarily awarded to the same laboratories each time, even if the laboratories are good performers.

The goals for the product and the standards to which they must be produced are not necessarily the ones that laboratory managers dictate. Customers define the quality requirements that are largely dictated by regulatory requirements and by government contract requirements.

Therefore, customers do not usually negotiate the requirements based on technical need, cost and quality with the laboratories. In some cases, the laboratory managers are forced to produce essentially the same type of data by using more than one process. This can be likened to having two separate and different production facilities produce the same product in a manufacturing firm. No one would chose to do this under normal circumstances. If two methods to perform the same test are maintained, it requires two distinct standard operating procedures, two methods that the operators must be trained to perform, two different sets of standard reference materials, two sets of inspection procedures, and two sets of paperwork procedures.

In addition, use of two or more sets of redundant procedures increases the probability for errors because the procedures are similar, but cannot be used interchangeably. The staff must make sure that samples are not inadvertently processed with an incorrect procedure, even though the final result will technically be the same. This is certainly not a cost effective process for laboratories. The laboratory managers are required to produce the data according to external specifications, even if there is margin to improve the process and product for customers.

At first glance, it would seem that the system was set up to purposefully stifle incentive for the producer of goods and services and assure that the supplier and the purchaser are at odds with each other. The laboratory managers must determine how to make a product that barely meets the standard requirements in order to bid in the competitive range for cost and still meet the minimum capability requirements to be competitive on a technical basis. If the laboratory managers fail to bid low enough to obtain a CLP contract, the laboratories lose the de facto certification that enables them to compete for additional business. (One must wonder, at this point, if the laboratory managers would compete for this type of work if there were a credible environmental laboratory accreditation or certification program that could be used in lieu of the de facto CLP certification.) The data purchasers feel justified in this approach because they are using good business sense to procure needed data at the lowest possible price. The problem results from a lack of perspective about the total process and the total cost of the process. It is no wonder that friction occurs frequently between laboratory data suppliers and the purchasers.

Originally, in the environmental laboratory business, the laboratories and the government worked cooperatively to jointly develop a means to rapidly analyze the great numbers of environmental samples that needed to be processed for the Superfund program. There was a spirit of cooperation and a "constancy of purpose towards improvement of product and service" that was of benefit to both the supplier and the producer. Where did that spirit of cooperation go? This spirit and the ways of the infant program were lost when the program grew to such an extent that the management of the process of acquisition of the analytical data became too large to allow interactive dialogue between the managers of the program and the laboratories that produce the data.

As the number of laboratories supplying data grew, so did the problems regarding comparability of data. Consistency in following the methods exactly became strictly enforced to assure comparability. Innovations and new ideas could not be implemented across the board. Therefore, the ideas took several years to implement across the program, or were never implemented because of the difficulty of changing equipment or processes in all laboratories. The many small changes and improvements that could lead to large differences in the data were not made. The management of the large numbers of laboratories, and management of the exhaustive inspection procedures that were developed to assure consistency of products from this diverse array of laboratories grew into a huge effort. As the program grew, innovations to the program and cooperation among technical, laboratory people and program managers diminished.

These problems are, in effect, the same problems that plague many large companies that have quality and production problems. The people on the factory floor have many good ideas on how to speed the process, eliminate waste, and improve products, but they are usually the last ones to be consulted for ideas. Some companies are turning this around and are improving productivity. This same change in attitude needs to occur with environmental data producers and purchasers — the laboratories and the those who are purchasing this data for decision making purposes.

2. Adopt the New Philosophy of Quality First and Participative Management.

The new philosophy is that quality is vital to survival of the company. The management must show through actions and words that quality comes first and that everyone in the company must work to achieve a quality product. Most quality problems cannot be solved without clear management action; this action must be taken to allow all the company to effectively work toward excellence.

Managers of the environmental laboratories are unable to effectively communicate to their companies that quality is of the utmost importance. These managers can communicate the message that the customer's needs will be met, but cannot say that this will ensure survival of the laboratories. The idea that quality products ensure the strength and survival of environmental laboratories that conduct a large percentage of their business with the government is simply not true. Environmental laboratories are required to produce products in a costly manner, and are not allowed to improve the production process to allow less costly production of the data. The idea that production of quality products in a quality process cost less is not allowed to flourish because the processes are prescribed and are costly if followed exactly. Environmental laboratories are forced to cut each corner possible on the prescribed process, even those that optimize the quality of the product, in order to make a profit. The Deming approach would be to optimize the process to produce the best possible product at the best possible price.

3. Eliminate the Need for Inspection by Building in Quality in the First Place. Cease Dependence on Mass Inspection.

The practice of inspection to achieve quality must be ended. Inspection at the end of a process to assure the product is the required quality is too late and is not reliable. It does not produce quality, at most it tells the purchaser only what the defects are. Quality must be designed into the process, or quality will never be achieved.

Environmental samples must usually be processed correctly the first time because there is normally only enough sample to perform the analysis once. The ability to perform rework on a defective product is not usually an option. In such cases that sufficient sample is available for reanalysis, the cost involved is usually the same as initial analysis. For these reasons, the analysis of environmental samples should be designed as a good process that minimizes causes of error.

Good production design and statistical process control are the solutions to minimize the use of inspection to determine conformance to quality requirements. However, based on the previous discussions about how production specifications in the environmental laboratory business have not necessarily improved quality, how can process specification and production design be the answer? There are several reasons that the current system does not work as well as it could. One main reason is that many of the steps and requirements in the production process do not necessarily add to the production of quality data. Many of the existing requirements, (especially if the variations in methodology that the analytical laboratories are required to perform are considered), might actually decrease the quality of the product. The optimum process is the one in which all steps are absolutely necessary, and all superfluous steps are eliminated. Superfluous steps increase the probability for error and therefore increase the inherent variability of the process. The object of optimum process design is to eliminate as many sources of variation as possible. The current process design has many steps that must be controlled. The need for each of the steps has not been sufficiently well assessed to determine if complete process steps could be eliminated in order to minimize errors in the process.

Another reason that the current system of process specification is an impediment to quality is because the process specified is not necessarily the best process and does not ensure production of a quality product. The statistical process requirements that must be met for a good product to be produced are not stipulated. The desired product is not defined; only the process that must be employed to produce it is defined. The process, as it is now defined, does not include a definition of the product at interim stages so that the process can be stopped before costly mistakes are produced.

An additional reason that the current process actually impedes quality is that inherent in the term quality is the term productivity. In order to truly be a quality process, the process must optimize production levels while, at the same time, it optimizes adherence to specifications. The current process specifications do not allow the environmental laboratory to optimize productivity by eliminating redundancy, increasing productive sample to QC sample ratios based on performance measures, or by any other cost-cutting and productivity improving measure. The production is strictly controlled by the process specifications.

The long term effect of strict adherence to process specifications, with no incentive for process improvement is the total stagnation of the improvement and development process. The effect of rewarding the industry for strict process compliance with little regard for the ramifications of step-by-step compliance with the quality of the final product causes a stagnation of innovation and creativity. If process control were balanced with appropriate product quality requirements, process requirements that made a difference to the quality of the final product would emerge as the important process control steps. The appropriate statistical process control specifications could then be determined because they would be based on the specifications for each step that are needed to produce quality products.

The effect of adherence to strict process specifications without appropriate regard to product quality requirements is to establish a "normalized" quality level for environmental analysis that is not necessarily the optimal level. The necessity to make a profit dictates a production line process in which the process controls are set at the most cost-effective level that will still meet the minimum specification requirements. The current practices almost guarantee less than optimal products will be produced.

The reader might question why the product specifications have not been historically more important than process specifications for environmental data acquisition. Usually, problems have occurred because of end-level inspection of the products, not because of inspection of the laboratory processes. This strange phenomenon is the result of the user of the product being unable to clearly define specifications for acceptable results of a chemical analysis. The purchasers of data could, however, define a process that should be capable of producing quality data.

The problem occurs in chemical sample analysis because the purchaser of data cannot know what analytes should be found in each sample. Therefore, it is difficult to determine if product data are correct. In order to ascertain if a process produces quality results, the purchaser needs to know what the results of the analysis should be. The purchaser has to know the range of acceptable values that the laboratory must produce. Since each environmental sample is different, the customer cannot determine from the "numbers" if the laboratory performs acceptably on the samples. The purchasers of the analyses have no measurement technique to compare the expected analytical results to those results that were produced by the laboratory.

When environmental laboratory programs were first designed, data purchasers had no recourse other than to define a rigorous process and define acceptable data as that produced by following the required procedures. The effect of this was to define an inspection procedure that did not inspect the final product, but inspected the documentation that indicated that the process was followed. This approach gives rise to an interesting problem. Now, the procedure for inspection of data at the CLP EPA headquarter's level is not inspection of a product, but is actually inspection to determine if a process is followed. This inspection is one step further removed from inspection of the end product and is not well correlated with the end product quality. To add to the problem, the EPA Regional data users many times inspect the data product — and have determined that compliance to process does not necessarily correlate well with compliance with expected product quality requirements.

This problem is not insurmountable. The purchaser can define the quality of analysis based on the product, or results, if there is an unknown QC sample included in each batch of analyses that can be inspected by the purchaser. Since the sample composition is known to the purchaser, but not the producer of the results, it can be used to determine if the laboratory performed an acceptable analysis on the sample. This can be used to infer whether the laboratory also performed well on the analyses of samples that were included with the unknown QC sample. The producer of the analysis could also include a sample as internal QC that is known by the laboratory to contain many analytes at set levels in order to gauge if their analysis was proceeding according to the specification. If the sample were used as a diagnostic sample, the laboratory could locate production problems before the analyses were actually completed. This would allow the laboratory to stop production when a problem was noted, not after the analyses were completed and all the samples processed incorrectly. More and better real-matrix type quality control samples to serve as statistical process and product inspection tools could alleviate this problem. Further, the current situation could be partially alleviated now by using currently available quality control samples.

4. End the Practice of Awarding Business on Price Tag Alone.

This statement is self-explanatory, but it needs to be expanded so the reader can understand the full intent of the statement and why the practice of awarding business on price tag alone is not a good idea.

Awarding business based primarily on price tag causes business to be awarded to the lowest bidder. If the quality of the product or service is not defined carefully, the result is the purchase of low quality products with the inevitable result of raising the total cost of the entire process. Instead of cutting cost by using the lowest bid process to purchase starting materials, management must strive to optimize cost by creating long-term relationships with single or fewer suppliers. Purchasing from a fewer number of good suppliers results in lower final cost for the purchasers and lower production cost for the suppliers. The end result is lower cost in the long term, or better quality products at the same cost, which benefit both purchasers and suppliers.

The practice of requesting bid prices on large purchases is common in both industrial and government acquisition processes. The intent is, of course, to minimize the cost of acquisitions. Vendors are usually selected based on price quotes to deliver products that meet certain specifications. Sometimes the bidder's facility and a sample of the product are also inspected to assure that minimum bidder requirements are met. Then, the bidder(s) that meet(s) the requirements and offer(s) the lowest price is selected. In many cases, several bidders are selected to provide the product.

There are several reasons that multiple vendors are selected. The first assumption in this decision is that all vendors selected will be able to meet the required specifications with little vendor-to-vendor variability. If variability is considered, it is expected that all vendors will be able to supply products whose variability is acceptable to the purchaser. The second assumption is that the practice of multiple vendor purchases is in the best interest of the purchaser because it heightens competition and therefore drives prices to their lowest possible level. It can also be assumed that the practice of multiple vendor purchases can also force improvements in the quality of the products so that the quality producers have a more competitive position. The third, and perhaps the most obvious reason for purchase from multiple vendors is that the purchaser fears that if there is only one source, or a very small number of sources, there is a risk that the supplier(s) will be unable to meet production demand for any number of reasons. The common supposed reasons for inability to support the purchaser include production failure, the company going out of business, staffing problems at the production facility, equipment failures, and natural disasters and other unavoidable problems.

At first analysis, it might seem illogical to argue with the three reasons that purchasing from multiple sources is usually chosen in preference to purchasing from one or few suppliers. However, this is not as good a practice as it might seem. Each of the three reasons will be examined in turn to determine if they are valid.

First, in order to make the assumption that all suppliers can meet total process variation requirements, the purchaser must know exactly what tolerances in the product are needed for use of the product. The purchaser must understand that variation is inherent in each vendor's process and in the products of each of the vendors. Hence, the more vendors making the products, the more variation one obtains in incoming products. If the purchaser strives to make the products as comparable as possible, minimal variation will never be achieved by purchasing from multiple sources. The purchaser must, therefore, determine carefully what level of comparability in products is needed and must rigorously inspect the incoming products to assure that these requirements are met. It is also implied that the purchaser has considered every type of variability that could affect product use. If one is purchasing from one or a few vendors and does not change vendors frequently, all usual sources of variation have probably been factored into the purchaser's specifications. However, each new vendor might have different sources of variation that could affect product use. Relying on many and different vendors mandates inspection to more requirements and might put the purchaser at risk with each new vendor of not inspecting for a source of variation that has not previously been a problem.

Since comparability of data is critical for environmental programs, it would seem to be prudent to eliminate as many sources of variability from the data is possible. One can argue that it can be important to have several suppliers so that all data for a study does not originate from a source that could have unforseen difficulties at time of data production. It can also be argued that the use of several suppliers will assure continuity and comparability of data from year to year on large monitoring projects. However, the acceptable variability of data must not be determined by using grand mean statistics of the variations in the data from all of the laboratories that are in the program. Acceptable variability must be determined by the comparability of results needed by the user of the data and must be based on how the data will be used. Analytical variability is acknowledged as a problem that affects the regulated community.[10] Since analytical variability can adversely affect companies that use data to support their compliance with regulations, it would seem prudent to minimize the variability that is inherent with using more than the fewest number of laboratories as is practical in any given circumstance.

Let us consider the second reason for purchasing from multiple sources. Competition among vendors is improved with a bidding process, unless quality vendors are driven out of business by vendors that barely meet the minimum specifications. In addition, the quality of products will only be increased by bidding and purchasing from multiple sources if quality products command fair prices and are recognized by the purchaser as superior to other products. In other words, if two firms bid competitively at the same price and one firm's products are significantly better upon examination of first and subsequent lots of product, the vendor should expect to preferentially receive more business because of the quality differences in the two vendor's products. If this does not happen, the incentive to pay close attention to detail cannot be expected to continue. The worst case scenario is that the poorer vendors are selected based on low price and barely meeting requirements and the better suppliers are driven out of business because they refuse to lower their prices to the point that a quality product cannot be made at a fair profit. In other cases, superior vendors become well enough known for the superior quality of their products and can sell to purchasers based on quality requirements rather than lowest-cost requirements. It is then not necessary for these vendors to bid for lowest-bid contract work, and they can refuse to make products for this type of customer. In such cases, the purchasers that use the lowest-bid procurement strategy will always be buying from lower quality suppliers that need any business that they can get.

Since the specifications to clearly define an acceptable product for the environmental analytical industry do not exist, the practice of underbidding and cutting corners to get a bid are rampant.[11] These practices make it difficult to complete and still produce only quality data. The current state of affairs makes it easy for the laboratories that produce shoddy work to compete and win business away from quality laboratories. This difficulty exists, in part, because reference materials that mimic the samples sufficiently well to use as indicators of sample analysis quality have not been readily available. This problem is receiving attention now, and should be at least partially remedied soon.[12] Further discussion of reference materials and their importance for successful implementation of both laboratory accreditation and data certification is included in Chapter Nine.

The third reason for using the practice of purchasing from multiple sources is the concern over the possibility of interruption of supply. This is a type of insurance premium that a purchaser pays to insure that the supply will not be cut off. How much do these "insurance premiums" cost? The purchasers should perform a careful analysis of the probability of supply interruptions based on experiences with good suppliers. Then the purchaser should determine if the interruption happened because the purchaser was "bumped" by a better purchaser that was more important to the supplier or because of a reason that could not be avoided by the purchaser or the supplier. If the supply interruptions happened because the purchaser did not do enough business with the supplier to be important to the supplier, this interruption could be eliminated if the supplier and purchaser had an important mutual business arrangement. If the interruptions could not be avoided, the cost of these interruptions should be compared in a cost and benefits analysis with the cost of maintaining several vendors, the added inspection cost associated with the practice of purchasing from multiple sources, the cost of inspection of several vendor's facilities, the cost of the added variability of the products and other related cost factors. In many cases, the cost of the problems that are encountered with poor suppliers are not reparable, because the reputation of the company using defective supplies suffers when poor quality materials are used in the final products. These expenses are in addition to the ones that are used for the cost and benefit analysis and should be factored in at the end of the analysis.

The risk of interruption in supply of analytical services could be mitigated by using EPA facilities to produce data, if necessary. In addition, the program does not need to eliminate all but one supplier, but lower the number of suppliers to a more reasonable number. In any case, the number of suppliers could be lowered and they could be asked to provide additional analyses if necessary. This will occur only if a cooperative supplier-purchaser spirit exists among the purchasers and suppliers of data. In addition, the suppliers would be less reluctant to analyze additional samples if they did not lose money on each additional analysis, as is now sometimes the case.

The three myths that exist about the benefits of purchasing from multiple sources have clearly been discounted in the previous discussions, but the benefits of purchasing from one or a small number of suppliers have not been fully established. Before the benefits are examined, it is important to take a different perspective of the production of a product than is usually taken. In order to understand the benefits of purchasing from a single source, one must look at the entire process that produces a product across company lines. Each step that is required to complete the product must be taken into account. If the producer of the final product needs incoming materials from a supplier to use in his production process, the supplier is considered an upstream extension of the production process. If the supplier of any product is then used to make another product, this supplier is simply a middle-man for the entire production effort.

The Deming way of describing optimal relationships between suppliers and vendors is that they cooperate as co-makers of the final products. This type of arrangement is beneficial to both suppliers and purchasers, including creation of a more effective working relationships for cooperative efforts to assure the production of quality final products. These rewards carry significant responsibilities for each party for making the co-maker relationship work.

The most important potential benefit of cooperative relationships between suppliers and purchasers is reduction in the variability of the materials coming into the process. This benefit reduces the scrap, rework, and need for adjustments in the production line caused by variation in incoming materials. Purchasing from a single source reduces this variability because the incoming materials from one supplier usually have less variation than the multiple types of variation that are seen if many vendors are used to supply incoming materials.

In addition, if the user of materials can identify the variations in incoming materials that can be eliminated by the vendor, the supplier and vendor can work together to find and eliminate these sources of variation. If multiple sources are used, the variations must be sorted and attributed to different vendors, and the purchaser must then work with the many different vendors to optimize their processes. If many different vendors are used, this is a significant effort. In addition, if the vendors are producing their materials for many suppliers, they will probably not want to invest the time and energy in the optimization of their process for one purchaser. If the vendor does produce most of its product for one purchaser, the vendor will optimize the product so that the major purchaser will continue with the business arrangement. In fact, if this vendor is economically dependent on a customer, each problem in non-conformance and variability of the product discovered by the major purchaser will be promptly addressed and resolved, if possible.

One can see that the use of fewer suppliers allows collaborative and prompt problem solving from the vendors and suppliers because production from one facility is interdependent on the other. In essence, the suppliers and vendors are working hand-in-hand to produce the product and the entire production process is optimized. This optimization of the process across corporate and government lines decreases the variability of the final product and therefore improves quality. Usually, a cost reduction in the overall cost of production also occurs.

What other benefits does purchasing from fewer suppliers have for the purchaser? It lowers the overall administrative cost of doing business. There are fewer contracts, fewer inspection problems, reduced paper work cost, and less time spent in negotiations with suppliers. In short, there is less non-productive overhead expenses and more time and energy spent on constructive improvements of the production process and product design. The headaches caused by ordering and inventory control, billing problems, and front-end planning for purchases can be alleviated. Companies can plan cooperatively to make sure that production of necessary supplies to meet the purchaser's demands are scheduled.

It is important to note that the single source approach can be further extended. The best approach for lowering overhead expenses for all purchases is to purchase as many incoming products as possible from the fewest number of vendors. In other words, the fewer the overall number of suppliers, the lower the overhead expenses. Therefore, if one supplier can supply other needed goods and services, consideration should be given to purchase of these supplies from that vendor because the cooperative arrangements are already in place with this vendor and the cooperative relationship can be further strengthened by purchase of a greater percentage of the company's production.

When the suppliers and vendors work together for their mutual benefit, there is a sense of confidence in the relationship that allows both parties to be creative and make corporate decisions that favor the long term health of the companies, rather than the short-term, bottom line. If the supplier knows that the purchaser will continue to purchase the starting materials from the vendor's firm, investments that improve the quality and reduce total cost can be made without fear of loss of the investment. Joint research and development can occur in which the engineers and designers from both companies can work together to design new products. The supplier's understanding about the supplier's product and the producers understanding about the final product and vice versa can be used to have a superior understanding about possible new products. Joint product development can occur only because there is a cooperative relationship between companies, rather than an adversarial relationship.

This point in Deming's quality management teachings is so important that highlights are reviewed here. Persons wanting to pursue this point further are encouraged to read an article on the subject that provided much of the insight presented here.[13]

The first assumption governing use of multiple sources is that if all suppliers produce materials that are within specification, then the variability between the incoming materials is not a problem. The fact is that even when all suppliers produce incoming products within specification, there are still variations in the incoming materials. Sources of variations include differences in equipment models, equipment maintenance, methods, training, operator skills, standard operating procedures, paperwork and computer processing techniques, management oversight, and quality assurance practices and inspection techniques. The cost involved in using materials with variation include increased rejects, increased rework, damage to production equipment, frustrated production workers who must constantly change the production process to accommodate variations, and the preferential use of one supplier over another in cases where this is possible.

The second assumption, that competition among many suppliers is healthy for competition and purchasing from a single supplier is somehow anti-free enterprise is not true if one looks at conducting competition with price as the major basis for the competition. When suppliers compete with each other on price, the price cutting might not occur because the production process has been improved and actual production expenses are decreased. In many cases, the cost cutting means cutting the corners in production of the materials by using cheaper materials or cutting production steps that are needed to assure a quality product is made. The final result could be a cheaper product at a lower cost. The effect of using cheaper products in the short term could be increased cost over the long term. Since each supplier is struggling with the pressures of making a profit, the strategy of providing just the level of quality that meets the minimal customer specifications will result. This results in products of marginal quality. Companies surely cannot afford or are little inclined to spend any monies on optimization of product quality when there is no incentive for this effort. Therefore, the long-term result is diminished quality of products and no ongoing research into product improvement.

The real competition in this country takes place between companies competing with other companies to serve as the sole-source suppliers to purchasers. If a quality supplier is edged out of one market by another supplier, then this supplier will compete to be the sole-source supplier for another purchaser in the same market. The result of vendor-purchaser cooperation results in survival of the best suppliers, production of quality products, reduction of overall cost to the customers, and a healthier national productivity. The result of this is a more competitive place for American made goods in the international marketplace.

The third assumption is that purchasing from multiple sources is needed to provide assurance that the supply of starting products will not be disrupted. This is extremely costly insurance. A cost and benefit analysis that considers the probability of disruption of supply and the cost of the insurance to assure constant supply will usually indicate that this practice does not make fiscal sense.

5. Constantly and Forever Improve the System of Production and Service.

Improvements to the production of goods and services must be made on a continual basis. Management must foster the goal that each and every person in the company must constantly seek new and better ways to make the product, process the paperwork, market ideas, and any other company effort that can be improved. Specifications must be continually revised to encourage improvement of products. Cost of manufacture and the time and effort required to produce a product or a service must be constantly minimized. Formal procedures that govern company production must reflect the company's state-of-the art knowledge and process control techniques. All company policies must be reexamined on a continual basis to make sure that they are not hindering the effort to improve quality.

The current means to purchase and supply a large part of environmental analytical data is through use of CLP-like contracts. This type of contract mechanism requires that the environmental laboratories perform the analyses by various, strictly-controlled methods. Improvement of methods is nearly impossible until after the contracts expire and other contracts are written. The contract methods spell out each step that is required for production of data. Any changes in the required process that decrease the overall cost of production are viewed as ways to make a higher profit and defraud the government, even if the changes do not affect the quality of the data.

This is not to say that the government and other large contract data purchasers do not request suggestions for improvements to contracts. These suggestions are solicited and many are accepted. However, these changes are implemented only when the contracts are revised prior to the bid process. Incorporation of small changes in a system that encompasses many contracts is difficult and costly. At present, in the CLP for instance, many government contracts would need modifications if a change were to be made in the methodology, reporting requirements, or other requirements. Clearly, the system is too large to allow small, but numerous, constant improvements. There is obviously little incentive to improve the procedures or the process in small ways that could immediately improve the system.

The current state of affairs in the environmental laboratory industry is clearly not in concert with Deming's fifth principle. The addition of small and numerous improvements from everyone creates the collective huge improvements. It is only when management fosters the collective incorporation of all improvements from each individual that total quality improvement is achieved. If management ignores opportunities for small improvements or does not immediately act on each good suggestion, the workers cease suggesting improvements. The workers do their job according to the unchanging procedures and produce products according to unchanging specifications. The constant improvement process is stymied and product improvement occurs at a snail's pace.

A deadly ramification of the environmental laboratory industry's efforts to improve production and lower expenses while still maintaining or improving the quality of the products is that such practices can be viewed as departures from the contract. Even if the final result of the changed and improved process is better data, the company can be investigated and be found guilty of not following procedures exactly as written. This can be considered waste, fraud, and abuse against the government. This is an important deterrent for companies to improve, or change in any way, processes performed under government contracts. There are, of course, instances when the changes are not in the best interest of a quality product, and the laboratories should be investigated for wrongdoing. However, the contract procedures that limit innovative changes to procedures also limit improvements.

6. Institute Modern Methods of Training on the Job.

This point is self-explanatory. Purchase of new equipment and improvement of production processes do not accomplish anything unless the workers are trained in their use. Quality begins with proper training. People cannot perform optimally unless they have a thorough understanding of their job, the equipment they use, and the quality requirements and techniques that are used to judge the quality of the work. Management must make sure that all barriers to using new equipment and procedures are removed. Management must also make sure that everyone understands quality assurance principles, the nature of variation, and simple statistics.

The current way of producing laboratory analysis fosters good training in-house. However, some positions require personnel qualification requirements that are stated as education and years of experience requirements. Because of these requirements, exceptional people that are well qualified for positions even if they do not meet the required years of experience requirement are held up because they do not qualify for some positions. The effect of such requirements is the ability to perform becomes less important than paper credentials. This problem can and does occur on a minor scale within companies even without the requirements of the national CLP contracts; however, the fostering of the practice on a national scale can cause severe damage to the entire laboratory industry.

Another problem that occurs because of the current way of doing business is that the sampling and analysis processes are totally distinct and segregated from each other. The sampler and the person performing the analysis might have no idea about the needs of the other person in dealing with the sample. In many cases, if the sampler had a better idea of the required analysis technique, the sampling technique used could enhance the quality of the analysis effort. In turn, if the person performing the analysis knew more about the sample and purpose of the data, the analytical technique employed could be complementary to the sampling technique. This cooperation could provide a better end product to the data user.

The person taking the sample, the person analyzing the sample, and the person using the data are required to be separated from each other by current CLP contracts. The understanding of the complete process and what is needed by the customer — or end user of the data — is usually a mystery to the persons producing the data. It is no wonder that the data user is many times disappointed with the results. The Deming's approach suggests that all individuals must thoroughly understand how their work effort impacts others and is interrelated to the production process as a whole. Clearly, this separation of efforts in the production of data does not promote total quality.

7. Institute Modern Methods of Supervision. Institute Leadership.

This point should be self-explanatory, but the authors fear that many readers might not clearly understand what is meant by leadership.

Leadership is facilitation. Management must learn how to facilitate the people's efforts to do a better job and make improvements. Managers must understand the production process and the statistical techniques that can be employed to identify what problems occur, and when they occur. It is the manager's job to improve the performance of people that are "out of statistical control" at the lower end and to use the skills of the people effectively that are "out of statistical control" on the upper end. It is the managers job to recognize one of Juran's teachings.[14] This specific teaching is that at least 80% of the failures in any organization are the fault of systems controlled by management. The focus of management is therefore on improvements to the systems so that the workers can do their jobs well.

Management in this country has been accomplished, for the most part, by the top-down management by control system. In this system, the top manager has certain goals for the year. These in turn, are translated into different goals at each level in the management chain. All the goals collectively add up to the top manager's goals being accomplished. For example, the head of a company could translate a goal for 25% increased profitability into a production manager's goal for increased production by 10%, increased sales for the sales manager of 20%, lower warranty expenses by the repair division by 50%, engineering design to decrease assembly cost by 10%, and so on. These goals then translate in lower management levels to sales quotas, rework goals, production targets, and other quotas.

This system is perceived to be working by management, at least on paper. There are, however, hidden negative ramifications of this system. Consider, for instance the manager who has to meet a production quota or not get a quarterly bonus. This individual could meet the quota by shipping many assemblies that were never put through final testing and inspection. This individual then receives the bonus because the production quota was met. The person that is required to lower the overall warranty expenses of the company bears the brunt of this production maneuver because the expense of rework will skyrocket based on shipping of untested assemblies. However, warranty expenses do not immediately go up. Therefore, the cause of the problem might never be linked to the production manager's quota and the shipping of untested assemblies. Warranty expenses might increase gradually as the equipment fails and is sent back for warranty work. The real culprit in this situation might never be found. In fact, the production manager might get promoted for his Herculean efforts to improve production.

Consider the effect of quotas on the production workers that have to meet the quotas for production of sub-assemblies early in the day. Once the quotas have been met, these people will not be rewarded for making more products. Because of this, these workers could stop work early and wait for quitting time. This is not the intent of the quota system, but the quota system fosters this type of mentality.

In most management by control systems, short term efforts are rewarded and the longer term efforts for improvement go unrecognized and unrewarded. When short term goals are unattainable, managers might skew the numbers so that on paper, the goals are met. This practice fosters secrecy and dishonesty. The managers sometimes begin to divide and start to "finger-point" at each other. People might "play it safe" so that they do not get caught taking a risk that someone can use as an example of a failure. A final point to be made is that management by control seeks to improve production and productivity by looking only inward and through belt-tightening. Very rarely do the top managers look to creative solutions that cannot be accomplished by simple goal and quota setting.

Managers must learn to manage and improve the processes by which work is completed rather than acting as enforcers of rules. They must focus on facilitating constant improvement and problem solving instead of blaming and controlling. They must look to the customer and the marketplace to assist in defining product quality standards instead of looking only within the walls of the organization. Processes must be optimized and people must be encouraged to make improvements in order for the company to improve the total organization by optimizing each step in the process and the ways that the processes interrelate with each other.

It is difficult for environmental laboratory managers to blame archaic management styles on the outside world. This is one Deming point that can be employed even if the outside world makes it difficult to optimize the benefits that could be accomplished through optimization of processes. The laboratory environment will be more productive if the managers are facilitators, even within "contract chemistry" constraints. Laboratory managers should strive to facilitate improvements even if the efforts are hampered by "contract chemistry" requirements.

8. Drive Out Fear so that All May Work Effectively.

This element is self-explanatory and refers to the fact that management should acknowledge responsibility for the majority of problems in the organization and work environment. Production workers must never be blamed for problems that are outside the realm of their power to fix. People must feel that they are secure in their jobs in order to ask questions or request help and make suggestions for ongoing improvement. Managers must be open-minded so that workers can question the effectiveness of current practices and suggest improvements. If fear of reprisal for suggesting changes to management occurs, constant improvement cannot take place.

Environmental laboratory managers have within their power the opportunity to implement all needed management changes to drive out fear in the laboratory only if they are allowed to implement constant improvements. However, laboratory managers must implement contractual procedures or request changes to these requirements. Recommendations for improvements to the contracts are usually not easily implemented. Therefore, improvements that would benefit laboratory operations are not usually implemented on a real time basis. Sometimes the over-eager laboratory manager that suggest too many changes can be viewed as a troublemaker to the government officials that oversee the contracts. The laboratory managers must therefore profess openness and co-operative problem solving in their own laboratory organizations, while they are constrained by contractual requirements in the types and numbers of improvements that they can implement.

There is little that laboratory managers can do to make huge changes that will improve productivity because of outside requirements imposed on laboratory management. The final effect of this situation might be loss of the managers' credibility with the staff. Many suggestions for improvement brought to these managers by their staff cannot be implemented because of contractual constraints. The end result might be diminished productivity and improvement in both individual laboratories and across the industry.

9. Break Down Barriers Between Departments or Staff Areas.

This point has been explained to some extent under the other points. Therefore, it will be only briefly discussed here. In essence, it is the responsibility of managers to break down the artificial division lines and follow processes through the invisible barriers of a company's organizations. The object is to improve product and process across company division lines. In order to accomplish this, each manager must view each process that affects the product or service as an internal supplier process. Each part of the process or service that is downstream to another process must be viewed as an internal customer. The object is to please the customer, whether internal or external. In turn, the supplier that provides the incoming product must optimize the product for the customer's specifications. All in all, each manager must have a good understanding of the entire company and the interrelated processes in order for optimization of quality in the entire company to be accomplished.

Individual environmental laboratory managers are at a distinct disadvantage to initiate significant process improvements since the procedures are strictly controlled by the purchasers. This has been stated before in this chapter. The constraints built into the contract procedures do not allow for complete optimization of processes. However, opportunities for effecting cost and production improvements in the laboratory exist. Laboratory managers must work harder to optimize and not change any contract requirements in the process. This is a difficult job, but it is not entirely impossible. The current contract structure makes it impossible to create optimal processes and makes laboratory managers' jobs more difficult. However, the manager must strive to make improvements despite these obstacles.

10. Eliminate Numeric Goals for the Work Force. Eliminate Slogans and Work Targets.

This idea was mentioned in the explanation of point seven. In the past, the use of slogans such as "Quality First" or "Buy American for Quality" without the underlying quality systems, quality design, and planning and resource commitments to effect quality of the products has been a great disservice to American workers. Workers have been told to produce better products without better tools, designs, or training. Management should not post slogans or insist on quotas if no concurrent effort is made to assure that these slogans will become reality and that constructive changes will allow quotas to be met. Even in the cases where many changes are made to allow improved production and improvements to quality so that quotas can be met, the quota system assures improvements that just meet the predetermined quota. It is better to enlist support of the company to make the largest improvement possible rather than set quotas. In this way, the organization works toward optimizing the processes to achieve the best production possible, which is better that striving towards an artificial quota.

The environmental laboratory manager is fully capable of motivating the work force without the use of slogans and work targets. In the current system, there are no slogans or numeric goals that are imposed from the outside on laboratories. Laboratory management must make sure that slogans and numeric goals are not employed because of the inherent problems that this would cause.

In addition, laboratory managers must lobby to prevent accreditation techniques that state that only "a certain number" of "a certain type of analyses" can be accomplished during a set time period by each person are not implemented. This type of system has been suggested in the past and is one that stresses numeric goals. In essence, it establishes numeric work goals for each person that performs a particular type of work. It hurts the people that take longer to accomplish tasks by rushing their work and judgement. It also penalizes the faster and more accomplished workers by requiring these people stop when the quota has been met even though they are capable of producing more results. The use of a work quota limits more capable individuals to the same compensation as the average worker because the benefits of their greater production capability cannot be realized.

11. Eliminate Work Standards and Numerical Quotas. Eliminate Quotas and Management by Objectives.

This element is similar to the preceding element. Quotas and goals address numbers and do not address quality and production processes. Quotas impair continuous improvement because the organization strives only to meet the established goals. Greater improvements are possible only without quotas.

Quotas sometimes have an opposite effect and decrease production. This is because unless each and every detail about a company's processes and products is specified by a quota or an objective, the organization might meet one quota and totally miss another. This occurs because their efforts are judged only on one area that contains a quota and not on the total responsibilities.

Meeting quotas without regard for the quality of product ensures inefficiency, high cost, and dissatisfaction from the customer. Management by objectives and the quota system promote arbitrary and short-term goals. The long-term goals must be the basis of decision making in order to assure optimum productivity and a competitive position.

Currently, some government-funded environmental analysis contracts require that laboratories accept a certain minimum number of samples per month if the government elects to ship samples to the laboratories. On one hand, the government does not guarantee that the quota will be used each time-period. In many cases, the quota of samples is not shipped to the laboratories for analysis. However, if the entire quota is shipped to the laboratory, the laboratory must accept the samples for analysis, even if the production facility has already scheduled other customers' samples. The laboratory must meet the production quota and cannot refuse to analyze the samples. This results in a hasty production job on the samples. The probable effect is lowered quality. If the laboratory were able to negotiate sample shipments that would assist them in normalizing their production, the effect would be a stable and normalized production process and heightened quality on all samples. This is not now an alternative for the environmental laboratories performing under some government contracts.

12. Remove Barriers that Hinder the Hourly Worker. Remove Barriers to Pride of Workmanship in Workers and Management.

This element is self-explanatory. Any and all barriers that prevent the worker from performing their best work must be eliminated. Every management system and standard operating procedure must be scrutinized to establish if it assists or hinders workers in performing quality work. Workers must be given the procedures, tools, training, and resources to perform well and produce excellent products so that they might have pride in their workmanship. The organization must not be forced to produce shoddy products in order to meet production goals.

Laboratory managers cannot blame any outside force for not implementing this element. The only outside roadblock to implementation of this element is that many required procedures might not be optimal. However, even with this constraint, laboratory managers can make sure that everything to assure that the workers can do a quality job is provided.

One other point bears mention here. It is currently an accepted procedure to underbid on certain contracts in order to buy "government CLP-de facto certification". If this is done, the laboratory management must develop a strategy to recoup the loss incurred by underbidding. Management must never force the workers to produce shoddy products in order to meet production goals that are set at inordinately high levels to recoup this loss. The analysts cannot be held responsible for recouping the losses by performing shoddy work on samples.

13. Institute a Vigorous Program of Education and Training. Institute Vigorous Program of Education and Self-Improvement.

All individuals in the organization must be trained in quality and their own respective disciplines. Managers must stress self-improvement as a means to advance in the organization. Each person in the organization should be encouraged to constantly learn new skills and forever improve his or her job skills. Such self-improvement should be used by the company to improve the processes, products, and services of the company.

The current climate in the environmental analysis business runs counter-current to constant improvement and self-improvement. The analysis by set production requirements makes innovation and creativity less positive attributes than they would be otherwise. People that are motivated and creative are the ones that consistently improve processes, products, and services. These persons are not as valuable in the current production of chemical analysis as they once were when constant improvements were valued and rapidly incorporated into sample analyses procedures.

14. Create a Situation in Top Management That Will Push Every Day on the Above Thirteen Points. Put Everyone to Work on Accomplishing the Transformation; Create Top Management Structure that will Push on other 13 Points Every Day.

The above thirteen points must be understood by management as an integrated set of management functions that equals a total quality management strategy when implemented in concert with each other. Each person in the organization must understand the points and strive to continually improve processes, products and services. Management must take the lead to implement each and every one of the thirteen elements on a continual basis.

Obviously, it is impossible for environmental laboratory managers to fully incorporate the thirteen points. It is impossible to fully implement total quality in environmental laboratory work given the current constraints caused by "contract chemistry."

TOTAL QUALITY MANAGEMENT AND ITS RELATION TO ACCREDITATION AND CERTIFICATION SYSTEMS

The last several pages have described the basic tenants of total quality management. The separate elements were presented and their application or lack of implementation in the environmental laboratory industry was explained. It should be obvious from the explanations of points 3 and 4 primarily, and the other points, that the vendor and supplier interface is critical to the effective implementation of total quality management. The supplier-purchaser interface is the point at which accreditation and certification systems exist. How can accreditation and certification systems improve vendor-purchaser relations and assist in facilitating trust and cooperation? The current systems usually cannot be applauded with fostering either trust or cooperation. The current systems usually are the cause of friction between the implementors of the systems and the organizations that are inspected.

Let us review the purposes of certification and accreditation. Most of these systems are designed and implemented to assure purchasers that supplier's organizations are capable of producing quality goods and services. The Contract Laboratory Program goes one step further and performs an inspection of product for conformance to process requirements. In essence, the certification and accreditation systems are designed to assure purchasers that suppliers are producing quality products based on established inspection procedures. The accreditation and certification programs do not foster much trust or cooperation because the inspection procedures might be seen as arbitrary and incapable of assessing true supplier performance. This is a usual effect of end-item inspection or spot-check assessments. Assessors always find something wrong, which is usually seen as a trivial item by suppliers. Suppliers answer predictable audit reports that do not usually address real production problems.

How can the accreditation or certification become more meaningful and foster cooperation and trust between suppliers and producers? This probably cannot be accomplished well in the absence of effective implementation of the entire Deming total quality management fourteen point strategy. Let us, however, seek to design and implement a system that includes as many of the fourteen points as possible, and encourages future implementation of all the points by designing into the system the flexibility to incorporate improvements.

THE PURPOSE OF CERTIFICATION AND ACCREDITATION IN RELATION TO VENDORS AND PURCHASERS

Let us reflect on the reasons for huge accreditation and certification programs. They are needed only if the purchaser does not know much about the supplier and the supplier's ability to perform. Either the purchaser does not need the service often, or uses so many suppliers that their work cannot be overseen effectively by the purchaser.

In attempting to understand the purchaser that needs the supplies often and uses many different suppliers, it is obvious that the purchaser does not understand the benefits of restricting the number of suppliers. This purchaser obviously does not worry about the variations in incoming data that several laboratories produce. This causes one to question if the purchaser knows what kind of variability is tolerable for decisions that the data will be used to support. It also causes one to question how the purchaser plans to inspect the data to determine if it conforms to these requirements in variability.

If the purchaser is using the certification and accreditation as the first level determination of the probability that a laboratory can perform quality work, this type of determination should not be needed often if the purchaser enters into long-term single or few-source arrangements with suppliers. Further, if the purchaser uses accreditation as proof of capability and follows up with more in-depth inspections of product variability and product conformance to requirements, this is a valid use of accreditation and certification.

Let us take this point one step further. The purchaser that uses accreditation and certification as a periodic verification of baseline competence, in lieu of such inspections by the purchaser is conserving QA resources. If these QA resources are used to perform in-depth verification of process and product specific to specific purchaser requirements, the use of existing accreditation and certification programs can lead to fewer redundant baseline capability inspections. QA resources could be used to establish ability of the suppliers to produce data according to purchaser's specific requirements. This would be a more effective use of QA resources because it would allow assessment of the process or products for conformance to purchaser's unique requirements. Since the suppliers and purchasers alike would benefit from an assessment of quality of products and services, the overall quality that the supplier provides to the purchaser could be improved. If this type of inspection were performed for a limited number of supplier-purchaser interfaces, the parties could work together to optimize total quality.

The insightful critic may now say, "I cannot accept the work of any baseline inspection system. I cannot trust any of the existing accreditation or certification systems now, so how could I possibly relegate my responsibilities to assure conformance with requirements to any such entity?" At first analysis, this argument has merit. However, if the in-depth oversight is provided with the newly freed QA resources, the next-level inspection by the purchaser provides some oversight of the inspection system. If the producer is incapable of producing quality products based on the in-depth assessment of ability that the purchaser accomplishes with the QA resources, this purchaser then has the facts to question the effectiveness of the baseline inspection system. This purchaser therefore has direct oversight of the inspection group.

In addition, it is common knowledge that many laboratories can pass baseline capability requirements, but fail to perform well. In essence, the ongoing in-depth assessment of products and services that will be provided by the myriad of purchasers that apply their QA resources to assess supplier's ability will provide real-time oversight of the baseline inspection system. If the purchaser has sufficient resources to inspect either at a baseline level or at an in-depth product and specific process level, the resources are better expended on a thorough product inspection process, coupled with information from an outside baseline inspection. A baseline inspection of capabilities program cannot assure the quality of purchased products and services, because the inspected suppliers still might be unable to produce what each purchaser requires. This must be verified by the purchasers.

In the previous paragraphs, the focus was on purchasers that had an ongoing need for suppliers' products. In this paragraph, the focus is on the purchaser that occasionally needs a product or service. These purchasers need an effective outside and impartial assessment of the quality of the suppliers' goods and services. The use of accreditation and certification systems is absolutely mandatory for this purchaser. The only type of accreditation or certification system that will provide this purchaser with any assurance that an organization can actually produce a quality product or service, is one that not only assesses baseline capability to perform, but also integrates product compliance to requirements in its final assessment. Thus, for both the small, occasional purchaser and the long-term, large purchaser, both assessments of capability and assessments of product conformance with requirements are needed. If one were to weigh the importance of the assessments, the assessments of products against product specifications would be more important than process assessments, because these assessments provide at least one-time verifications that the processes work.

VENDOR ACCREDITATION AND VENDOR CERTIFICATION — ARE THESE SYSTEMS NEEDED?

It is strange that so many baseline inspection programs exist and so few product compliance programs exist. The only rational explanation for this phenomenon is that the oversight authorities that implement the baseline inspection programs do not have the resources to implement the next level of inspection, that is, the in-depth process and product inspections. If all the QA resources were not expended on baseline capability inspections, one would assume that the more detailed inspections could be implemented. If one views accreditation as the baseline inspection for capability, that is a pre-qualification requirement, then many of the problems that are now encountered with accreditation programs are not issues. The problem with the current accreditation programs might not be with the programs themselves, but with how they are used by purchasers. Baseline capability inspections do not guarantee good products and services will be produced by the accredited organization. Yet, this type of guarantee is what most purchasers expect from accreditation programs - even their own. This is a misuse of the programs. It is no wonder that accreditation programs are not viewed as satisfactory to purchasers.

Only the purchasers of goods and services know exactly what process and product specifications are important to the end use of the purchased goods and services. Therefore, it is important that the appropriate specifications are determined and communicated to the producers before purchasing goods and services. Only in this way are the purchasers' needs conveyed to the producers. The suppliers cannot be held accountable for manufacture of goods according to specific requirements unless those requirements and the specifications for those requirements are fully communicated to the suppliers. Only the purchasers' inspectors or agents, or the suppliers' inspectors, or the two inspectors working in conjunction with each other can provide evidence that products are made according to specifications.

A third-party certification authority can assess supplier inspection records and products to determine compliance of products to product specifications. Likewise, the processes can be certified with analogous evidence and inspections. If several purchasers have exactly the same purchase requirements, the possibility exists for third-party certification of processes and products according to product specification requirements. This is not strictly in keeping with Deming's principle on use of single suppliers, but does improve vendor-supplier relations by providing working requirements and set specifications. This facilitates a good working relationship between vendors and suppliers since compliance to specifications is judged by a third party. Using a third party also decreases the possibility that purchasers can change their interpretations of requirements and that different inspectors can issue different interpretations. Third party inspection programs require standard interpretations and do not allow interpretations of specifications to change. If a third party observes on-line process and product inspections that take place at the vendor's facility and then certifies the products and processes with no redundant inspection and testing, the best use of QA resources is accomplished. The goal is to eliminate wasted steps and to make the best use of QA resources. Complete elimination of redundant and costly audits, inspections, tests, and other review activities that add no quality improvement to the products and processes is the goal.

The current trend toward international quality requirements for products makes it imperative that vendors and purchasers determine consistent sets of requirements and specifications. Only by specifying recognized international requirements will the quality of American goods and services be able to conform to international requirements. In addition, it is imperative that the laboratories and testing and inspection authorities employ methods that are consistent with international requirements to judge conformance of products to international requirements. The most critical baseline requirement upon which all other inspections and testing is based is the competence of testing authorities and inspection agents. There is clearly a need for accreditation of baseline capability of these agents. There is also a need for further assessment of testing data by the industries or their agents to assess the quality of the data produced by the testing laboratories.

IMPROVEMENT OF CURRENT PROGRAMS THROUGH APPLICATION OF TOTAL QUALITY MANAGEMENT PRINCIPLES

The pessimist could say that the concepts of total quality management will never be fully implemented in this country. Rather than argue about what the future might hold, it should be determined how best to implement quality management given the apparent constraints and eliminate as many constraints as possible. The fourteen Deming principles are not now fully implemented. It must therefore be determined how best to work with the systems that exist and slowly change them into systems that support total quality management.

Baseline capability inspections, that is, laboratory accreditation programs can be used to further implement the total quality management strategy if properly used by purchasers. Vendor certification programs that assist vendors and suppliers in eliminating redundant inspection procedures also can promote the total quality management strategy. The existing systems can be improved to become support systems for total quality management by eliminating redundancy and becoming a part of a credible national program. Since they are cost-effective and efficient QA systems, they will promote total quality management in the laboratory industry.

REFERENCES

1. Wettach, Robert, "Function or Focus?- The Old and the New Views of Quality," *Quality Progress* 18(11): 65–68 (1985).
2. Joiner, Brian L. and Peter R. Scholtes, "The Quality Manager's New Job," *Quality Progress* 19(10): 52-56 (1986).
3. "Quality Systems Terminology," American National Standard ANSI/ASQC Standard A3-1987 (Milwaukee, WI: American Society for Quality Control, 1987).
4. Sullivan, Edward and Joshua Hammond, *On a Silver Platter* (Milwaukee, WI, American Society for Quality Control, 1986), p. 11.
5. Deming, W. Edwards, *Out of Crisis* (Cambridge, MA: MIT Center for Advanced Engineering Study, l986).
6. Juran, J. M., Editor in Chief and Frank M. Gryna, Assoc. Editor, *Juran's Quality Control Handbook*, Fourth Edition (McGraw-Hill Book Company 1988).
7. Ishikawa, Kaoru, *Guide to Quality Control*. Asian Productivity Organization. Publication obtainable from the American Society for Quality Control, Milwaukee, WI, (1976).
8. Stanenas, Al P., The Metamorphosis of the Quality Function," *Quality Progress*, pp. 30–33 (November 1987).
9. Langevin, Roger A., "SPC and the DOD," *Quality Progress* 21(9): 35–37 (1988).
10. Koorse, Steven, "False Positives, Detection Limits, and Other Laboratory Imperfections: The Regulatory Implications," *Environmental Law Reporter*, pp. 10211-10222 (May 1989).
11. Felsen, Harvey G., "What's Wrong With Our Industry?" *Environmental Testing Advocate* II(1): 2 (1990).
12. "Summary of the Reference (RM) Certification Group Meeting of Tuesday, May 8, 1990," Memorandum to Reference Materials Task Group, A2LA Accredited Environmental Laboratories, A2LA, Environmental Assessors, A2LA Environmental Advisory Committee, Reference Material Suppliers, A2LA Board of Directors and Other Interested Parties, Memorandum distributed by the American Association for Laboratory Accreditation, Gaithersburg, MD (May 23, 1990).
13. "Deming's Point Four: A Study," in *Quality Progress* (December 1988) pp. 31-35 from Ranney, Gipsie, and Ben Carlson, Ed. (Vandalia, OH:The Ohio Quality and Productivity Forum, 1988).
14. Juran, J. M., Editor in Chief and Frank M. Gryna, Assoc. Editor., *Juran's Quality Control Handbook*, Fourth Edition (McGraw-Hill Book Company 1988) p. 17.6.

Chapter 7

THE SOLUTION — A SYSTEM FOR INDEPENDENT THIRD-PARTY CERTIFICATION OF ANALYTICAL DATA

DESCRIPTION

During the course of our previous discussion, we have highlighted the chaotic state of the present system of laboratory accreditation and touched on several of the more serious short-comings associated with the present system. We have also alluded to a more logical approach — that of an independent third party accreditation system. We have additionally discussed the third party approach to product certification as embodied, for example, by the Underwriters Laboratories and the National Sanitation Foundation. At this juncture, we will present a generalized approach to the third party process.

Accepting the foregoing premises, what are the major attributes of a successful accreditation system? These can be categorized into two major areas:

1. Appraisal of the capability of the laboratory to perform
2. Performance evaluation and ongoing monitoring of the laboratory's actual performance and ability

Performance evaluation and monitoring are related to defining the laboratory's ability to produce analytical data of a quality sufficient to meet the requirements of the data's end use. However, these procedures are usually not fully implemented in conjunction with baseline capability assessments. Therefore, the usefulness of most accreditation programs is usually limited. In order to fully explain the need for both types of appraisal, these two major areas will be examined in detail.

APPRAISAL OF PROCESS CAPABILITY

The capability of a laboratory to satisfactorily perform any type of testing involves a number of generic criteria. The laboratory evaluation process must consider:

1. Laboratory organization — Is the organization of the laboratory designed to ensure the effective and efficient generation of analytical data?
2. Personnel — Are they adequately trained and experienced? Are there sufficient personnel to accomplish the work?
3. Facilities — Are these adequate to accomplish the work in a safe consistent manner?
4. Equipment — Is the equipment adequate to perform the indicated analyses and is it maintained in a manner to guarantee accurate reproducible results?
5. SOP's — Are the procedures and methodology sufficient to accomplish the work and are they maintained in an orderly, updated manner?
6. Reference Standards and Performance Evaluation Materials — Are the materials of sufficient quality, stored in a manner that preserves their integrity, and used in accordance with procedures that minimize laboratory errors that occur from incorrect preparation and use?
7. Chain-of-Custody — Are the procedures for handling samples sufficient to establish that the integrity of samples is not compromised and that analysis results are correctly assigned to corresponding samples?
8. Documentation Procedures and Records Retention — Are the data maintained in a manner that will allow easy access to historical data, including raw data if it is needed? Are retention requirements reasonable and clearly stated.
9. Data Reporting Format and Content — Are the data reported in a manner that allows the purchaser to determine the quality and usefulness of the analysis.

10. QA Plan — Is the laboratory QA plan adequate to provide quality
     oversight during the entire process of data generation or simply a
     check of the final product?

It is interesting to note that this list of items that must be inspected
to determine baseline capability is similar to the description of the CLP
on-site inspection described in Chapter Five. However, there are
several additions beyond the CLP list to make the inspection of
capability complete. These items are not included in the CLP
assessment because the program requires specific data deliverables,
provides standard reference materials, and does not require the
laboratory to retain documentation and records. Because of this, the
CLP does not inspect to requirements that are not relevant to the
program. Therefore, the CLP can provide a blueprint for an
accreditation and certification program, but must be scrutinized to
determine where requirements must be added and deleted.

APPRAISAL OF THE ABILITY OF THE LABORATORY TO PERFORM

Use of Performance Evaluation Samples to Assess Laboratory Ability

Even if baseline capability is determined, some assessment of ability
is necessary to provide evidence that the laboratories can produce
acceptable results. Without this evidence, the accreditation is severely
limited. This assessment of ability is similar to product certification,
since both process and product are assessed.

For this reason, perhaps the most important factor in the process of
evaluating a laboratory's ability is the performance evaluation or
proficiency testing sample(s). These materials are designed to provide
an objective evaluation of the laboratories' ability to perform the
analytical test or methods being evaluated.

The best assessment of ability to perform occurs if the performance
evaluation samples are like the samples that the laboratory will be
required to analyze. For example, if the organization requesting the
accreditation evaluation wishes to have the laboratory perform the GC/
MS analysis of a selected series of volatile and solvent extractable
organic components from a waste stream, the proficiency testing

sample(s) should consist of known amounts of the analytes of interest in the matrix of interest. In a like manner, if analysis of selected metals and metalloids in a sludge is of interest, the proficiency testing samples would consist of the metals and metalloids in the proper matrix.

Admittedly, this is a straightforward concept and is certainly obtainable. However, this straightforward approach has never been completely instituted and remains a goal seemingly at odds with reality. Yet, the benefits to be gained from these "real world" proficiency samples are enormous. Current efforts by several standards suppliers[1] to prepare and distribute proficiency samples could help to mitigate the limited use of performance evaluation samples. Current efforts[2] to provide for certification of the performance evaluation and reference materials will promote the manufacture of reliable materials that can be used by industry and government to measure laboratory proficiency.

If a sample of a well-matched matrix to the samples to be analyzed cannot be obtained, the laboratory can be challenged with a synthetic sample that contains the analytes that are analyzed by the method that will be used. This is not the best type of performance evaluation sample, but it does provide a measure of ability to perform. These samples should be utilized when matrix matched sample are not available.

The results of the laboratories' analysis of these proficiency samples — assuming the "true" values for the analytes of interest are known — provide an objective benchmark on which to evaluate the laboratories' abilities. These performance evaluation materials can also be an integral part of other important aspects of laboratory performance, such as QA audits and data review.

Use of Performance Evaluation Samples to Evaluate Data and Perform Laboratory Monitoring — Measures of Ability

Performance evaluation samples can provide a snapshot appraisal of the processes and data produced by laboratories at a specific time. These samples should also be used as the basis of an ongoing procedure to determine the quality of data on a day-to-day basis. An ongoing appraisal of the data from samples whose analyte composition is known is necessary to determine day-to-day performance.

Critics of performance evaluation programs often state that the laboratories' performance on quarterly or yearly samples are the laboratories' best efforts and are not an adequate appraisal of day-to-day analyses. These critics are correct in many cases. Few laboratories actually analyze the samples as routine analyses. However, the performance evaluation samples do alert the accreditation body of laboratories that are incapable of performing even with their best effort. This is valuable information.

However, because the only fair assessment of ongoing performance is a system that provides ongoing appraisal, the use of performance evaluation samples must be expanded. Proficiency testing samples can be used to monitor laboratories' performance. These samples can be placed blind into the laboratory in all or periodic lots of samples being analyzed. The results of these analyses will serve as an instant check on laboratories' performance. This information can be assimilated and used as an integral part of the evaluation of the laboratories' capability to perform.

The next step of the process is to define how to use the data in a formal laboratory evaluation process. These samples must be used on a continual basis by either the accreditation body or the purchasers of laboratory data. The data from these samples must be used with the baseline accreditation to provide evidence of continual quality performance or alert the purchaser of the data that the results might be questionable.

Use of Performance Evaluation Samples for Appraisal of the Ability of the Laboratory — Data Assessment and Certification

As an unbiased benchmark, the data from the analysis of proficiency testing samples can be useful as the basis for conducting QA audits of data. These audits should serve as a means for reviewing the entire operation of the laboratory as it was operated during the time that the data were produced. Since inspection of the data allows the purchaser of data or the accreditation authority to assess actual ability, complete data must be delivered with the testing report in order to complete this type of assessment. Therefore the accreditation authority and purchaser need to specify exactly what is needed to complete this audit and require this data to be delivered. Items or procedures to be reviewed include, but are not limited to:

1. Sample receipt, storage, and tracking — as evidenced by the data and supporting documentation that accompanies the data
2. Equipment operation and maintenance — as evidenced by the

supporting documentation and data and the actual results of the analysis

3. Qualifications, experience and training of personnel — as evidenced from the supporting documentation, results on the sample, and the interpretation of the data itself

4. Adherence to the QA Plan and SOP's — as evidenced from the documentation providing evidence that procedures and analytical methods were correctly followed

5. Data inspection system — as evidenced from the inclusion of faulty data, omission of required documents, and other errors that the laboratories' internal inspection should have found and corrected before reporting the data

6. Laboratory facilities — as evidenced by the inclusion of contamination, degradation of standards, and other problems that occur if the facilities are inappropriate or not maintained properly

7. Adherence to appropriate good laboratory management practices — as evidenced by the delivery or non-delivery of quality data

By using the results of the analysis of proficiency testing samples as a guide, the auditors will be able to determine at what point mistakes, if any, were made. As a consequence, the evaluation of the laboratories' ability will be a reasonable in-depth assessment. This information will be extremely valuable to the organization requesting the accreditation and certification and to the laboratory itself. This information can provide the basis for problem resolution and efforts to correct deficiencies.

Review of analytical data on a periodic basis is an extremely important aspect of performance evaluation. This data review can be designed to encompass the wide variety of data types. For instance, data destined for use in litigation efforts might require complex and extensive reviews. Data for use in areas of a less critical nature might require a less extensive review. The end use of the data will dictate the review process.

It is important to note at this point that the data audit must be directly related to the purchaser's use of the data for the audit to be most useful. An accreditation body's assessment of data to requirements that are not necessary for specific data users has limited usefulness. This type of audit, however, is useful to determine if the laboratory is able to meet data requirements that were specified. The result of the audit provides an indication of ability, even if they are not the most relevant to the data purchaser.

In order to conduct the in-depth, third-party reviews of data that are appropriate to the purchaser's use of the data, quality specifications and requirements and the associated data certification criteria must be determined. These specifications and requirements must be based on actual data user needs. In that way, the data can be produced in a cost effective manner and meet the data user's needs. Only when the requirements are fully delineated, can an audit that is relevant to the data user's needs be conducted by a data purchaser or a third-party auditor.

Regardless of the extent of the review, the results of analysis of proficiency testing sample(s) serves as a basis for defining the quality of the data. As such, the data review can be designed around the end use of the data rather than forcing all data into a single format based on the most stringent end use.

The data evaluation can be designed to assess the use of the data and provide an assessment of the laboratory system that produced the data. This on-going assessment of data and actual laboratory performance will surpass any snapshot performance evaluation and scheduled on-site laboratory evaluation visits.

Such a monitoring program presupposes that performance evaluation samples will be available and that the third party accreditation organization is capable of providing these materials and maintaining a secure data base on laboratory performance. All aspects are obtainable and such a system will virtually ensure the credibility and wide acceptance of the third party approach. The use and availability of performance evaluation materials is critical to this program. This subject is covered in detail in Chapter 9.

SPECIFICATIONS AND REQUIREMENTS OF THE SYSTEM

Capability and ability must both be assessed for any accreditation body to provide meaningful information to data purchasers. These two requirements can be separated into two requirements:

1. Laboratory Accreditation — The assessment and determination of baseline capability.
2. Data Certification — The assessment and determination of ability as evidenced by inspection and certification of product — that is the performance evaluation samples and data from the laboratories.

These two requirements must be carefully integrated into a single program to provide the data purchasers the necessary indication of a credible system. In any evaluation process, the most important single aspect is the competence and integrity of the evaluators. Without competent and principled evaluators, no accreditation or certification process will be perceived as credible or have widespread acceptance. Recognizing this gives rise to several critical specifications and requirements for the successful functioning of a third party accreditation process:

First, the third party must be independent of management constraints and undue pressure, from both the requesting organization or agency and the laboratory community. This does not imply lack of accountability; the contrary is true. The third party is accountable for providing a credible, technically competent evaluation to the organization requesting the accreditation. It also is responsible to the laboratory community to provide a technically competent, unbiased, evaluation of the laboratories.

Second, in order to accomplish this task, well trained and highly motivated evaluators are needed. The complexity of the evaluation process for both "capability" and "ability" aspects demands that the evaluators be trained in all facets of the evaluation process and that these individuals maintain an up-to-date understanding of the laboratory industry and methodologies.

Technical competence of personnel is essential to any successful accreditation or certification process. As has been mentioned previously, the perception of lack of competence in current accreditation systems is real and significant. This fact has been a detriment to all accreditation and certification programs. Only by hiring and retaining competent personnel who are allowed to maintain their technical competence can any evaluation or accreditation organization hope to be successful. Only by compensating these people with commensurate wages will accreditation and certification systems attract this talent. Perhaps by consolidating redundant systems, the effective buying power of the one system will be large enough to attract these skilled individuals.

Many laboratories fear being assessed by professionals who might gain inside knowledge of particularly effective procedures during laboratory inspections only to leave the assessor's organization to join a competing laboratory. If fewer accrediting systems existed and the systems were able to compensate skilled assessors, the systems might be better able to retain the assessors and minimize their changing positions and moving into positions in the laboratories that they once evaluated. Since this fear is prevalent and could impact the types of positions that assessors could be able to hold if they chose to leave the accrediting organization it is critical to compensate these professionals fairly. It is important to stress that the assessors must not be allowed to divulge any trade secrets, information about competitive edge processes, or any confidential information that is obtained through their role as assessors.

Third, the third party must be capable of interfacing not only with its client organizations but also with the laboratory community and the scientific community as a whole. This will require not only technical competence but also strong management acumen and personnel with a high degree of interpersonal motivation. This ability to interface successfully with a wide array of organizations will go far toward assuring independence and acceptance.

By successfully initiating the above requirements, the third party can be assured of successfully meeting the specifications required for a successful accreditation process. These criteria will always be designed to answer the "capability" and "ability" questions. Moreover, the role of the accrediting organization can be expanded at the option of their client organizations, to include technical assistance to the laboratories in the form of problem resolution and corrective action.

THE ROLE OF ACCREDITATION AND CERTIFICATION BODIES AND THE ROLE OF DATA PURCHASERS

An accreditation or certification program will not ever assess all data for data purchasers. Data purchasers must determine when assessment of data is necessary and obtain these services. Third party organizations can perform these services for the data purchasers. Unfortunately, data interpretation is usually performed solely by the purchaser of the data when it would be best performed in close concert with the data producers. Currently a significant weakness exists in data interpretation in many regulatory organizations and regulated industries because the skills needed to perform this service are unique and are not widespread. A third party could interpret the data and work with both supplier and purchaser in an unbiased role. The third party could determine, in many cases, how the data could be used even if the data were not perfect. The third party would not be providing the service to benefit the purchaser or the buyer, but rather to derive the most information content possible from the data. The results of the assessment could serve to dissipate some disagreements between the buyers and purchasers of data. Since the intent of the interpretation would not be to benefit either party, the third-party interpretation would, of course, be unbiased and therefore important to the use of the data.

The main point here, is that the concept of an independent third party organization is capable of being expanded to encompass many aspects of data generation and utilization. Third–party organizations can be used to provide both accreditation and certification services. The most critical aspects to be maintained for both inspection of process for capability (accreditation) and inspection of products for conformance to specifications (certification) are competence and integrity.

THIRD-PARTY CERTIFICATION AUTHORITIES

The need for an accreditation system that is credible, widely accepted, and national in scope is critical. This is not simply the opinion of the authors, but as has been previously discussed, this is the prevailing opinion of many people in regulatory agencies, much of the laboratory community, and has been discussed by the scientific community at large.[3-5] Granted that this concept is widely espoused and is a logical approach to bring order out of the total chaos that is inherent in the present system of laboratory accreditation, what third party organization exists to fill this vital niche? Unfortunately, no organization currently is capable of providing this service in its entirety.

This is not meant to imply that existing organizations cannot provide this service. It states that, at the present time, no organization fulfills all the essential requirements. What is required is an organization that has a history of independence, technical competence, credibility, and above all, integrity. In addition, the organization should have a history of implementing similar baseline process capability (accreditation) and ability (product certification) programs. With this as a basis for development, the appropriate third party organization or organizations can be developed that will be capable of providing accreditation and certification services that will be widely accepted and highly credible.

The authority to perform this evaluation/accreditation service ultimately resides with the third party's client organizations. Recognizing that the ultimate accountability must always be the purview of the client organization, the third party accreditation and certification organizations function as an extension of this authority. It is incumbent upon the third-party organizations to perform the accreditation and certification processes in the most competent, unbiased manner possible both to ensure the third-party organizations' credibility and to show the client organizations that the assessments have validity.

While there are currently no organizations totally geared to provide this service, organizations such as the National Sanitation Foundation and Underwriters Laboratory are two examples that currently have a history of experience in providing product certification services. These independent third–party organizations have a track record of technical expertise, acceptance, and credibility that can be used as a basis for developing the required evaluation and certification system. There also exist accreditation bodies such as the American Association for Laboratory Accreditation (A2LA) that can fill the niche of providing capability evaluations that could be followed with product certification programs. It remains only for the concerned parties to institute efforts to move toward a logical and workable approach to laboratory accreditation and data certification that may include these organizations. We can only hope that reason prevails in the near future and all affected parties work collectively to meet that goal.

OVERSIGHT OF THIRD-PARTY ORGANIZATIONS

It is important to remember that the accreditation of capability and certification of data solution is different from current accreditation processes. One problem with current systems is that there is little ability of the parties using the results of the accreditation bodies to measure the competence of staff and the actual merit of the accreditations. In this system, the clients of the accreditation and certification system will have two indications of laboratory ability and capability — not just one as with current systems. The laboratories will be inspected to determine capability. Further, they will also be continually monitored to establish on-going ability, based on both performance evaluation samples and periodic data audits. If the laboratories cannot perform on performance evaluation samples or on data audits, the clients can quickly determine that the third-party accreditation service is performing an inadequate job, or that the accreditation requirements are not appropriate for the specific type of data the client is procuring. In any event, the client has direct oversight of the third-party assessment organization because of the information provided from the performance evaluation samples and data audits. The best oversight is provided if individual clients include performance evaluation samples to provide additional ability assessments and also conduct additional data audits. The use of performance evaluation samples for this purpose is expanded in Chapter 9.

REFERENCES

1. Gaskill, Alvia, "Environmental Matrix Standards: Benefits and Limitations," *Environmental Lab* 2(2): 10-12 (1990).
2. "Summary of the Reference (RM) Certification Group Meeting of Tuesday, May 8, 1990," Memorandum to Reference Materials Task Group, A2LA Accredited Environmental Laboratories, A2LA, Environmental Assessors, A2LA Environmental Advisory Committee, Reference Material Suppliers, A2LA Board of Directors and Other Interested Parties, Memorandum distributed by the American Association for Laboratory Accreditation, Gaithersburg, MD (May 23, 1990).
3. Keith, Lawrence H., "Guest Editor Reply to Letter to the Editor," *Environmental Science and Technology* 23(1): 6 (1989).
4. Miller, Wade, "Conference Summary, First International Conference of the International Association of Environmental Testing Laboratories," International Association of Environmental Testing Laboratories, Arlington, VA, November 21, 1988.
5. "Availability, Adequacy, and Comparability of Testing Procedures for the Analysis of Pollutants Established Under Section 304(h) of the Federal Water Pollution Control Act, Report to the Committee on Public Works and Transportation of the House of Representatives, EPA/600/9-87/030, pp. 1.8–1.9 (1988).

Chapter 8

IMPLEMENTATION CONSIDERATIONS

We like a man to come right out and say what he thinks — if we agree with him.

Mark Twain

Perhaps the major issue preventing implementation is agreement on what is needed. Until all key players can agree on a solution, the current status of laboratory accreditation and certification will remain. All parties that will be affected by a national accreditation and certification program must begin to communicate so that a solution can be agreed upon and implemented.

INTRODUCTION

This chapter provides the framework for implementation of a national accreditation and certification system. It is not the intent to provide all the answers, but simply to provide a framework that all concerned parties can use and expand upon in working towards a fully implemented national system.

Besides providing a framework to implement the system, this chapter includes the answers to certain key questions. Who should act so that the system becomes fully implemented and is maintained as a credible system in the future is suggested. Also, the entities that might accept the accreditation program as valid are identified. Many of the implementation problems are difficult to address. Discussions of some of these problems that are expected to be encountered in the implementation effort include several viable options and include the pros and cons of the options. A specific issue section at the end of the chapter is provided to address issues of concern that have not been included into the discussions of this chapter, but still should be addressed because they add to the information known about implementation of similar systems.

In the last chapter, the detailed steps for accreditation and certification and the necessary traits of an independent third-party were addressed. This chapter provides more detail on how these concepts can be crafted into a system that can be implemented for many different types of laboratories and can provide for the different needs of the purchasers and suppliers of data.

ROADBLOCKS TO IMPLEMENTATION OF A NATIONAL PROGRAM

The previous chapters have provided the reasons why a national accreditation program for laboratory "capability" is necessary. They have also provided an explanation of the differences between laboratory accreditation and data certification. Further, the need for both the accreditation and certification parts of a national laboratory assessment program has been presented.

If the need for the combined accreditation and certification system seems so clear, why has a national system not been created? The previous chapters have alluded to many reasons. This chapter will delineate these reasons and provide some logical approaches for removing the roadblocks that stand in the way of implementation.

The current state of laboratory accreditation and certification is chaotic at the least and detrimental at the worst. The excessive numbers of formalized systems — not to mention the almost limitless types of informal systems — has created a situation in which excessive duplication of effort and unnecessary cost are normal. Still new piecemeal systems are currently being proposed, mandated, and implemented. This lamentable situation results in increasing duplication of effort, increasing cost to both purchasers and suppliers of laboratory services, and diminishes certainty regarding the quality of analytical data.

Rather than increase the number of accreditation systems ad infinitum, perhaps a more successful and assuredly a more rational approach would be to reduce the number of accreditation systems. Concomitant with this reduction in the number of accreditation systems, a reduction in the number of accrediting entities that provide redundant services would occur. Moreover, duplication of effort would be greatly reduced or even disappear. Further, the cost of laboratory services, in both time and money would also be reduced. Considering the obvious benefits to all concerned parties, who could be against a consolidation of accreditation systems? Sadly, nearly everyone involved with laboratory accreditation has a reluctance to fully accepting a national accreditation program.

RELUCTANCE OF THE FEDERAL GOVERNMENT

A crucial roadblock to remove is the federal government's reluctance to fully acknowledge the need for a national system. The authors do not intend to criticize the federal government for this reluctance. The purpose of discussing this roadblock is to address the reasons why it exists. The reasons are complex and must be discussed in order to understand the federal government's perspective of the situation. Only then can implementation alternatives be drafted that can be acceptable to the federal government.

One obvious concern of the federal government is the cost and other resources that would be necessary to establish and administer a national program. This could be a powerful roadblock. However, the federal government does not need to bear the burden of the cost and administration of an accreditation program if the government acknowledges the need. The federal government only needs to acknowledge and use a national accreditation and certification program if the program meets the federal government's needs. If the government chooses to use the accreditation program as a baseline program that could be augmented if necessary with specific federal or state requirements, the federal government could be able to realize substantial cost savings in the reduction of the number of current baseline assessment programs.

Reduction in the number of these systems could mean that the current burden and cost of administering several systems could better be spent in a cooperative system, without loss of final authority of the government systems in determining compliance with federal and state requirements. The overall system could be designed to preserve the necessary autonomy of different programs and provide for their special needs. However, the redundant aspects of all the programs could be effectively reduced by a cooperative venture. No manager would lose responsibility for the assessment of their program's data. Each manager would maintain responsibility, but be able to use the existing QA resources to accomplish this job more effectively. The existing QA resources could be used more effectively by these managers to assess special, not baseline, requirements. These QA resources could be used for program-specific performance evaluation samples or data audits, for instance.

This roadblock can be removed with the advent of an acceptable solution. Until now, no one has provided reasonable alternatives and acceptable solutions. In order for any solution to work, the federal government must be willing to cooperate in efforts to improve the current situation. The government might cooperate by first acknowledging the possibility for change and then by assisting in the planning, implementation, and oversight of a national system. The government needs to provide input so that the system is acceptable to their needs. It is important for industry and government to openly communicate their concerns about accreditation and certification and discuss the real reasons for these concerns so that progress can occur.

The EPA has already taken steps to investigate the possible need for a national accreditation/certification program.[1] Other government agencies such as USATHAMA have also initiated longer term efforts to discuss this possibility.[2] These efforts should be applauded and the private sector should assist those government entities that have been charged with studying the problem. It is hoped that the EPA group will request assistance from industry so that the effort will progress quickly and be cooperative with other similar private sector efforts.[3–5]

The political ramifications and turf battles inherent in such a consolidation effort could be terrible to behold. Without exception, government departments and agencies (federal, state, and local) will be reluctant to accept a truly independent accrediting organization. Much of this reluctance resides in the area of "comfort" factors. In other words, the evil you know is better than the evil you do not know. Another cause of reluctance is a fear of losing control. Yet, these concerns should not pose an insurmountable obstacle to a successful third-party accreditation system. The utilization of independent third parties as laboratory accreditation and data certification organizations ensures more comprehensive, technically competent oversight of laboratory capability and ability. It also removes the government from the untenable position of applying the seal of approval to the work that they are responsible for. Simply put, independent third parties depend on credibility and accountability for their very survival.

More germane to the discussion of potential obstacles are the areas of widespread regulatory acceptance and government purchasing requirements. Federal agencies, such as EPA, FDA, USATHAMA, accredit and certify laboratories. These are sometimes called certification systems, such as EPA's Drinking Water Certification Program, or they might be de facto certification systems, such as EPA's Contract Laboratory Program. If the government agencies that administer these programs could accept a third-party system for the appropriate part of their baseline assessments, the system's credibility would be assured. Acceptance by the various regulatory bodies within governmental organizations will make not only the concept of independent third-party accreditation legitimate, but will also ensure its acceptance throughout the laboratory community.

All the listed agencies and numerous others wish to maintain their respective oversight programs. However, government agencies are increasingly dependent on contractor support to accomplish their missions. The merits of using contractor support versus completely in-house programs can be debated ad infinitum. However, the simple fact remains — the widespread use of contractor support to accomplish laboratory oversight has resulted in these skills being located in the private sector. Utilization of contractor support in this area has led to the perception that the government is not technically capable of providing laboratory oversight. This is really not the problem. The problem is determining which way laboratory oversight can be accomplished and what will be the most effective use of both government and private sector talent.

Government personnel must maintain the critical jobs of regulatory and policy promulgation. The skills needed to perform this work are certainly different from the technical laboratory assessment skills needed to monitor laboratories. Government personnel cannot maintain all skills needed to perform all jobs well. Most government personnel that have been involved in laboratory assessment in the past have needed to be experts in all areas and have not been given the opportunity to remain technically up-to-date to perform laboratory assessments. Since they are no longer functioning as technical staff, this would not be cost effective. In addition, their real functions are to manage programs and not perform analyses themselves. They simply have too much to do to remain competent at the bench level.

In using contractor staff to perform periodic scheduled laboratory evaluations, there is also a widely held perception of conflict of interest. Many government and laboratory personnel believe that contractors are intent on expanding their support base, even at the cost of inflicting unnecessary hardship on the laboratory community. While not totally true in either case, enough truth could be inherent in these perceptions that they cannot be ignored. The solution to the dilemma of whether to use individuals contracted by the government to perform laboratory capability assessments or whether to perform the work in-house by government staff is solved by using an independent third-party accreditation system as a baseline assessment. It is to be hoped that reason will prevail and that the various regulatory organizations will embrace the concept of the independent third party.

If this were to occur, the resources now expended in the government's laboratory evaluation programs could be better utilized with the production of more and better performance evaluation materials, better remedial on-site evaluations, more method evaluation studies, and many other services that will not be provided unless additional resources are available.

RELUCTANCE OF LABORATORIES TO ACCEPT A NATIONAL PROGRAM

Laboratories are reluctant to accept one national program that will replace existing programs. At first glance, this might seem strange. However, upon further analysis, this reluctance is easily explained.

The current myriad of systems makes it possible for a laboratory to always be accredited by some system. A laboratory can continue to do business with most clients even if it loses one type of accreditation or certification because there are so many different certifications and accreditations. If there were one system, the possibility of losing all business because of loss of accreditation would be a real concern. If the one national accreditation authority and its processes were entirely flawless and provided a real-time assessment of all laboratory ability, a laboratory that always produced quality work would not fear loss of business based on laboratory accreditation. If the accreditation authority only stopped sample flow for the exact sample types and for the exact time frame that the laboratory was experiencing difficulty, the competent laboratory would have no fears of the accreditation authority. This is because during times when the a competent laboratory is out-of-control, these laboratories are not producing data for clients. They are attempting to resolve their problems before processing samples. If an accreditation body could catch the laboratories that were producing poor results and cease the appropriate sample flow for these periods of non-performance, the laboratories that produce quality results would benefit. In addition, data purchasers would benefit.

Obviously, an accreditation process will never be totally perfect and will never provide real-time assessments of all data. The possible flaws in a single national system and their potential ramifications on the laboratories' business must be considered. Solutions for dealing with the possible problems must be worked out far in advance of their occurrence. Only when potential problems with the national accreditation program are resolved will laboratories fully accept the idea. It is important to point out that many laboratories are eager to resolve such issues so that a national program can exist. This is evidenced by several efforts to promote this idea. The International Association of Environmental Testing Laboratories (IAETL) Task Group on Laboratory Accreditation,[6] the American Council of Independent Laboratories' (ACIL) Consortium for Quality Environmental Data,[7] and the long effort of the American Association for Laboratory Accreditation (A2LA)[8] to promote a national program are examples of this effort.

One other reason for some laboratories' reluctance to accept a national accreditation and certification program is because it would make it easier for all laboratories to perform testing in different states. Some of the artificial business barriers that are caused by state certifications and state fees would disappear if a national accreditation and certification system were accepted by the states. Therefore, competition from out-of-state laboratories would occur to a greater extent than it currently does. The current systems reduce this national competition. Therefore, many small, local laboratories are against a national system and many large, national companies endorse the concept of a uniform national system.

An additional reason for some laboratories' reluctance to accept a national system is because several large laboratories have expended large amounts of time and money in gaining many certifications and accreditations to provide evidence of competence to their clients. The implementation of a national system would diminish the importance of this past investment.

Finally, a fear that some laboratories have is that the national accreditation or certification program that will be implemented will be as weak as many current systems. They do not want to support an effort that will not provide a high level of assurance to data purchasers that data quality is acceptable.

THE GENERAL APPROACH — A SCHEMATIC DESIGN

A national, unified accreditation system is feasible. Such a system can be designed to accommodate the needs and requirements of the broad spectrum of laboratories within the laboratory community and the government and industrial purchasers of data. A generalized approach to such a system is presented in Figure 1. It is important to clearly note the links that hold the players together. This general approach will be discussed according to the players, their actions, responsibilities and the integration between the players and their actions.

While this scheme is both generic and uncomplicated, it embodies the germane features of a national independent third-party accreditation process. In essence, the organizations (government agencies, industry, etc.) provide the acceptance criteria. The accrediting organization(s) evaluate the laboratories, based on the criteria and report the results to the requesting organization(s). If desired, the requesting organization(s) can review the evaluations and accept or reject accreditation for individual laboratories.

Does this approach negate the independence of the third party? Absolutely not. The recommendations of the third party are independent of the requesting organization. However, the final acceptance of the laboratory performance evaluation is the purview of the requesting organization. Final accountability resides with the regulating body. If the requesting organization delegates this accountability to the third-party accrediting organization, they must do so with full knowledge of what they are delegating. The merits of this delegation can be positive and should be strongly considered by the requesting organization.

Recognizing the generalized nature of the accrediting process, are we truly improving the efficiency of the process? The present approach to laboratory accreditation embodies the generic aspects necessary to any accreditation process. In addition, the requesting organization can choose to define more specific criteria relevant to particular types of laboratory services. It is important to remember that additional specific criteria are not always necessary or desirable. The capability and ability evaluations will often be sufficient to define the accreditation process.

As such, the simplification of the accreditation process will result in a greatly improved efficiency and effectiveness.

ACCREDITATION AND CERTIFICATION CRITERIA

In order for the accreditation body to make an accreditation determination, it is necessary to evaluate at least two different sources of information. One source provides the information that is used to indicate the laboratory's capability to perform. The other source of information provides evidence of the laboratory's ability to perform. These two sources of information have been described in Chapter 7.

TIERED APPROACH TO DEFINING REQUIREMENTS

We have listed two sources of information needed for determination of accreditation and certification. The requirements and the assessment of the laboratories in relation to these requirements is critical. We must recall that the objective of accreditation is to determine if the laboratory is producing data that is suitable for use. Therefore, there will be as many different types of requirements for data as there are uses of data. Will this result in a confusion of matrices versus requirements and subtle differentiations between acceptable laboratory data and unacceptable laboratory data? It can result in a hopeless state of confusion if it is not designed correctly, but this does not need to occur.

It is essential to the success of the accreditation process to determine what is critical to the determination of capability and ability on a generic level. It is then necessary to define a tiered approach for determinations of capability and ability for more specific levels. A natural consequence of the overall simplification of the accreditation process is a tiered approach for evaluation of laboratory capability and ability criteria. As delineated in the approach to accreditation presented above, not all laboratory services require equivalent evaluation to be "accredited." For instance, all requesting organizations will have situations when evaluation of the laboratory's baseline and generic capability and ability will suffice to define a baseline for accreditation. In such cases, specific criteria are unnecessary. Moreover, within the capability and ability criteria, not all criterion need to be universally applied.

One area where requirements will likely differ is in the frequency and

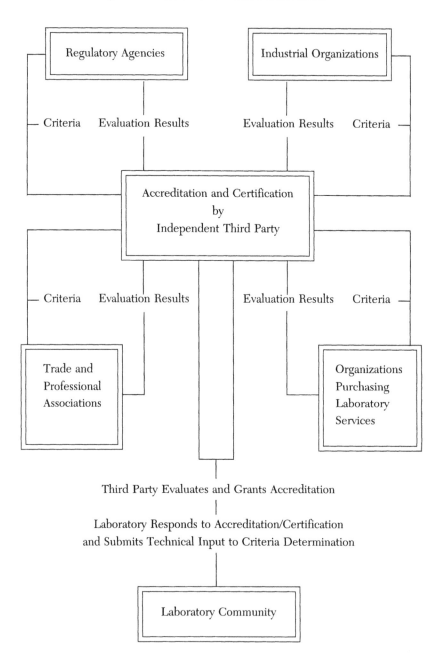

FIGURE 1. The general design — a schematic approach.

type of data review that is considered necessary. It is unlikely that total data review on all data will be an accreditation requirement for non-regulatory functions. The requirements for data review will vary significantly because the need for this monitoring will be determined in several ways. For example, the need for data review could be determined, in part, by the use of the data and, in part, by past performance of the laboratories involved. A perfect example of differing data review needs based on data use is data collected during the remediation phase of site clean up conducted by Superfund. When excavation operations are in progress, the need for analytical data is critical. However the data does not need to be of an accuracy and precision to three decimal places. The data needs only to be of sufficient quality so decisions regarding continued excavation can be made. Obviously, following completion of the clean up, the data used to demonstrate the success of the remediation must be of suitable quality to make critical decisions. The data review procedures used in these two situations would certainly be different. In addition, the methods used to obtain the data could also be different.

If a tiered process of laboratory evaluation and data assessment is employed, one critical design element of the system will be to differentiate between the tiers. This must be performed well so that all laboratories and data users understand what each tier of accreditation and certification means. The system must be logically designed and well presented so that the lowest tiers of accreditation and certification are well differentiated from higher tiers. The key to a successful tier system is education of data users to what extent the limitations and representations of both ability and capability are delineated by the different tiers.

THE FUTURE SYSTEM — QUALITY ASSURANCE OR CHAOS?

Now that we have stated that many differences in data, from the analytical methods used to process samples to the data evaluation procedures used to assess data can and should be different in the future, one must wonder how anyone can call this effective quality assurance. After all, the assurance of quality in environmental data has, in the past, required all aspects of the analysis and review of data to be the same. In this way, comparability of data from all laboratories was accomplished.

This is a valid concern. However, we must remember that this system will provide a baseline that should be augmented with assessment of data for specific uses. This should have been accomplished on all past data. The future system is based on what has been learned from the past.[9] Before the advent of highly structured contractual systems for obtaining data, data was produced to support the decisions at hand. The errors, usually reported as precision and bias, were obtained for each data set. The point at which the analytes could be quantitated was determined for each data set. Quality control samples were run along with the samples, if possible, to provide a measure of performance.

The CLP contract procedures and methods provide huge amounts of data based on specific requirements. However, the program realizes that without some data assessment, the data cannot be effectively used. The baseline capability and ability assessments can be routine. However, these baseline assessments do not necessarily assure production of comparable and useable data. For this reason, careful monitoring of data and efforts to minimize variability between data suppliers is important. The purchasers of the data must augment the baseline system and also assess data to determine if it is useable for decision making.

When the CLP was developed, much was not known about environmental measurements, capabilities of laboratories to perform these analyses, data assessment procedures, performance evaluation samples and all other aspects of environmental measurements. However, the experience of environmental chemists and the myriad of data produced from these analyses now allows industry and government to take a routine, but more appropriate approach to quality assurance. Therefore this system is based on past QA and QC principles that evaluate each data set, and more recent large program QA and QC systems.

In this system, the data generated must be compared to a baseline that is established by the intended use of the data, which may not necessarily be the routine CLP requirements. The accreditation and certification systems assist in establishing baseline capability and ability. The more in-depth assessments by data users determine data useability for particular purposes and are tailored for specific data user needs. These detailed assessments according to data user needs are the final tiers of accreditation and certification. Some interim tiers can be provided by a third-party accreditation and certification program. Final tiers that are additional to any tier of a national program will be provided by the data users or their agents.

"NEW" REQUIREMENTS

The Common Technical Baseline — Performance Evaluation Samples

Most of the new requirements that will be needed are centered around evaluation of ability, since most other accreditation requirements were centered around evaluation of capability. The most critical of these are

- Expanded use of reference materials to establish comparability and for data evaluation and ongoing monitoring
- Use of common data reporting, storing and retrieval standards to facilitate the use and evaluation of data

The first new requirement listed is needed to provide comparability between data and the assessment of the methodologies on a real-time basis. Tantamount to the success of this new approach is the use of good reference materials. Chapter 9 provides more detail about the use of these materials. Correct use of the materials will allow data sets that are produced by different laboratories, by different methods, and at different times to be utilized on the same study.

Currently, the use of various sets of data that have been compiled from different places in different times are difficult to use for making decisions. In the future, every effort possible should be made to allow environmental data collected to be potentially useful for all environmental studies. It is unfortunate to see the battles among even federal agencies in their attempts to establish method and QA equivalency so that the data can be used for different purposes. The problem is that it has been virtually impossible to determine what effect the different required processes or methods have on the data. Perhaps one agency was correct on one point and the other agency was correct on another. It was impossible to establish correctness. A system that requires that the reference materials analyzed by the method of choice provide evidence that the quality of data needed for the decision, as described by precision, bias, and level of quantitation, can alleviate this unfortunate situation.

This solution could facilitate cooperation among the various federal agencies that have vehemently disagreed in the past about these subjects. All the agencies have disagreed because they have expended tremendous efforts developing their respective systems and believe that their systems are valid. By using reference materials as the common ground among these systems, the best of all of them should be evident. The differences between these systems should cease to be important, and the best features of all the systems can be merged in a cooperative effort.

The goal for the future is to produce all environmental data in a system that is defined and acknowledged by all to be valid. In this way, all environmental data will have the potential of being used by any decision maker that needs the data. The decision maker will be able to evaluate the data's usefulness based on its precision, bias and other such information. Some decisions might require collection of additional data, but the data that is already collected can provide valuable information about the exact type of additional data needed so that unnecessary sampling and analysis is avoided.

In addition, the object of environmental analysis is to improve the overall habitability of our environment. In order to achieve this goal, it is imperative that the data collected for each media be merged with other information to provide an overall picture of where and for what reasons human and ecological risks are the highest. In addition, data about soil, polluting industries, groundwater tables, surface water flow, weather patterns, geological data, and many other sources of data should be merged for use in clean-up planning, emergency preparedness planning, and other uses.

The Common Information Management Baseline — Standard Data Reporting Format

It is not acceptable to continue to collect environmental data that can only be used for one purpose, and to collect it in such a way that the results are not comparable with other sets of data. The time has come for EPA, environmental offices within other agencies, and industry to cooperate and define the baseline rules, data reporting formats, and other requirements that will facilitate the collection and use of comparable data. A national accreditation program is a first step in this direction. If all environmental laboratories are assessed by the same system, their data reporting requirements and formats must, by necessity, be standardized to facilitate data review and assessment. This standardization will facilitate the collective use of data for more purposes.

The key requirements needed to facilitate information management are

- A standardized way of reporting all data that includes key standardized sample and quality control data elements.
- Data reporting specifications that are oriented to facilitate data use for the primary user.
- Data reporting specifications that are organized to facilitate general use of data by all future secondary users.
- Specific processes for storing and retrieving raw data, including QC data and documentation.

These same reporting requirements will facilitate the use of the reported data and the stored raw data for data assessment by both data purchasers and the accrediting and certifying body(s) at either established or random intervals. The need for all data purchasers to have the raw data at their access at any future time is paramount for its use. This will also facilitate future, secondary uses of data.

OLD REQUIREMENTS — REDESIGNED

Redesign of On-Site Procedures

Most of the old or classic laboratory accreditation procedures were centered around capability indicators such as evaluation of information provided to the accrediting authority on applications. The applications usually included information about personnel, facilities, equipment, SOPs, and the facility QA Plan. In addition, many of the accreditation programs perform on-site verifications of the applicant's information. This baseline inspection is sometimes repeated at a set time frequency. In other cases, site visits are conducted because of technical considerations, change in laboratory facilities or location, or a change in the types of samples that the laboratory analyzes.

The on-site inspection has usually been a combination of a checklist and comments that support the checklist findings. The on-site visits are conducted in this manner to facilitate the equitable assessment of all laboratories by comparing them with set checklist requirements. In addition, the checklist approach facilitates the use of the results for comparative evaluation of all laboratories and automated data storage. The checklist also allows people with limited assessor experience to perform approximately the same evaluation as more experienced evaluators. This last benefit of the checklist approach is a questionable benefit, because the inexperienced evaluator should not be assigned to independently assess laboratories. However, partly because of the lack of experienced personnel to perform assessments, the use of the checklist approach has been mandatory in most programs.

The use of unstructured assessments in which the evaluator conducts the assessment by asking open ended questions allows a skilled evaluator to gain a more in-depth understanding of the laboratory's capabilities. However, it allows the process to go out-of-control if the assessor does not conduct the evaluation professionally. In addition, there can be great differences in the content of the assessor's on-site visit reports. There is the potential of conducting vastly different assessments at different facilities in which the same aspects of the laboratory are not evaluated the same by different assessors, and the necessary baseline aspects of laboratory capability are not obtained. The checklist approach limits the possibilities of these types of problems.

In redesigning the baseline capability assessment process in a national accreditation program, it is necessary to use experience to structure the best capability assessments for the future. Several examinations of the requirements of the many environmental programs have been conducted.[10] An especially thorough study was sponsored by the state of New York and Touche Ross and Company and was conducted by the Dynamic Corporation in 1988.[11] This report, entitled "Comparison of the New York Environmental Laboratory Approval Program (ELAP) with Other Quality Assurance Programs" indicates that there are many aspects of baseline capability assessment that are the same throughout environmental accreditation programs. The application forms and assessment checklists for all programs can be merged into a baseline application and capability assessment that will provide information to all parties. In addition, the application form and checklist can include the needs of separate organizations. The on-site inspections can gather this information for utilization by the appropriate parties. If this information is not relevant to all parties, it need not be used by them. In addition, the extra evaluation cost associated with the more detailed assessment could be effectively subsidized by the organizations needing this extra information.

There is no reason that the application and on-site assessments could not be tiered to allow baseline information and any additional tiers that are necessary for separate organizations. In addition, if the requirements of individual organizations are so special that a separate application and on-site evaluations are necessary, these separate applications and visits can be more effective because they do not need to repeat the baseline assessments, but can be limited to the extra requirements. In this manner, the resources for these activities that are important to only special data purchasers could receive the resources they need to be quality assessments, rather than added-on to the current procedures. Currently, the added-on approach expends resources ineffectively by assessing all capability indicators, and does not do an effective job of assessing special needs. The resources now expended could be more effectively used by developing and using special performance evaluation materials and tools that are specific to special areas of need.

Addition of Non-Traditional Methods and Procedures

An important challenge to be faced will be to expand evaluation efforts to the non-traditional environmental data collection areas. Historically, only environmental data and laboratories that produced specific and limited types of environmental analyses were assessed by such programs as the CLP, USATHAMA, and Drinking Water Certification Programs. Other environmental laboratories and methods, such as mixed-waste procedures, hazardous waste engineering treatability assessment methods, field analytical/screening chemical methods, soil porosity, asbestos methods, toxicity and other non-traditional methods have not been included in any current environmental laboratory and data capability and ability assessment programs. It will become increasingly important to develop the means to assess capability and to include non-traditional data in standard reporting formats so that these types of data can be better assessed and used for environmental decision making.

The focus of initial efforts to design a national system will be to consolidate and redesign traditional methodology assessment programs. However, non-traditional methodology, equipment, personnel capabilities, and QC techniques must be included in the planned tiered structure. These assessments for non-traditional environmental testing will be fully developed and implemented in the future. If they are now included in the accreditation structure, they can easily be implemented in the future.

As environmental chemistry techniques expand and steps are taken to more effectively use data from non-traditional methods, the accreditation and certification tiers for non-traditional evaluation will become increasingly important. Since this system will be designed around baseline capability coupled with assessment of performance based on performance evaluation sample analysis and audits of data, separate evaluations based on different performance evaluation samples and data audits will be accomplished without restructuring the system to add non-traditional methods. As performance evaluation samples become more available to facilitate the use of non-traditional methods, these same materials will allow for assessment of ability of laboratories to perform the methods.

It is critical to include the non-traditional methods in the master design. In the future, use of these methods could surpass current routine methods. As environmental professionals become more sophisticated in the use of biomonitoring methods,[12] future data on separation and identification of chemical analytes by traditional means could become a less important part of the environmental data base and the biomonitoring methods could become more important.

More In-Depth Assessment of Internal QA and Self-Inspection Systems

In any accreditation and certification assessment process, more emphasis must be placed on the actual quality assurance practices and procedures followed in the laboratory. If the process works and the inspection procedures are closely followed, the laboratory might be capable of catching out-of-control situations so that poor data are never delivered to customers. Current assessment of capability procedures look only at the parts and not the whole laboratory process and its capability to produce a quality product. This change in the focus of on-site inspections can be accomplished by a more process oriented approach, including a more process oriented checklist and open-ended questions that track actual laboratory processes through sample acceptance, analysis, and reporting.

RESPONSIBILITY FOR DEVELOPMENT OF QUALITY SPECIFICATIONS AND REQUIREMENTS

Considering the overall generalization that data are of a quality sufficient to satisfy the final use specifications, accreditation criteria can be developed to cover a wide array of situations. In the proposed third-party system, the accrediting and certifying organization(s) would share the responsibility for defining appropriate criteria with the requesting organization. This "tiered" approach has the inherent flexibility to address all aspects of laboratory accreditation across the grid of laboratory types because baseline and special requirements can be drafted to meet a wide array of needs. The baseline requirements define a baseline accreditation. Additional assessments to special requirements can provide information for specific data purchasers in addition to the baseline assessment and can be termed a different type(s) of accreditation(s).

The accrediting organization must be fully accountable to those organizations or groups utilizing its expertise. To accomplish this, there must be extensive communication between the requesting organization and the third-party and between the third-party and the laboratory being accredited. This is particularly critical in situations where evaluation criteria are "tiered" to meet the final use of the analytical data.

Alluded to in the preceding discussion is the concept of data quality or perhaps more appropriately, quality specifications. Central to this concept is the acceptance and understanding that quality can have various acceptable levels and these levels can be defined with sufficient detail that evaluation criteria can be developed around these levels of quality. As practicing analysts, we might be loath to subscribe to this concept. After all, we always produce the "highest quality" data. This is a commendable idea, but it is more rhetoric than fact. In truth, data will never be totally equivalent and most assuredly are not all the "highest quality." This does not imply fraud, incompetence or shoddy workmanship. For example, thin layer chromatography (TLC) data are of a lesser quality than quantitative gas chromatography/mass spectrometry (GC/MS) data in terms of the data's usefulness in the identification of many different analytes. This is not to say that TLC is less useful for specific applications or that GC/MS is always the only answer. That would obviously be absurd. What it does imply, is simply this — the quality of data can be specified based on use requirements and therefore, the useability of data can be delineated and specified. In a similar manner, quality specification criteria and requirements for laboratory accreditation can be defined and specified.

Who is responsible for defining and providing these quality specifications and requirements? The independent third party? The third party certainly has a high stake in this process. Perhaps the requesting organization? Again, they must be considered to have a great deal of interest in the quality specifications and requirements. What about the laboratory community? Does it have a role in defining quality specifications and requirements or would their participation be like letting the wolf guard the sheep?

The most reasonable answer is that all concerned parties have a vested interest in defining quality specifications and requirements for accreditation criteria. In the instance of the third-party accreditation organization, it is obvious that they must have a dedicated interest in defining the accreditation criteria. After all, their existence depends on the wide acceptance and credibility of their accreditation. Consequently, the third party must be highly motivated, technically competent, above reproach and unbiased.

As with the third-party accreditation organization, the requesting organizations existence depends on obtaining data of a quality sufficient to satisfy the end use of such data. This interdependence between the third-party accreditation organization and the requesting organization is a mutually beneficial, symbiotic relationship and is indispensable for the effective definition of accreditation criteria.

This interaction will ensure that all pertinent areas of concern are adequately addressed in the accreditation criteria. Further, the experience of the third party will greatly benefit the requesting organization in widely diverse areas. Paramount among these areas are technical expertise, the capability to interface with a broad spectrum of organizations, and techniques to reduce redundancy and greatly increase efficiency and effectiveness. As these groups of organizations gain experience in the proposed accreditation process, the efficiency, effectiveness and acceptability will increase. This process is not self-limiting and can only increase the credibility of the proposed accreditation process.

"But what of Barabus?" Is it possible that a widespread interest in quality specifications and requirements could exist in the laboratory community? The answer is a resounding YES! Obviously the laboratory community as a whole has an interest in determining the next accreditation hurdle they must jump. But more than this, the vast majority of testing laboratories have a desire to produce useful data. Indeed, sad to say, rather than the regulatory agencies, it is often the testing laboratories that provide the impetus for improvements in analytical method performance and data quality. Accepting that the continued existence of independent laboratories depends on appropriate data quality, the laboratory community also has a vested interest in defining widely acceptable accreditation criteria.

Recognizing the interdependent nature of the accreditation organization, the requesting organization, and the laboratory community, it is apparent that all three groups have a vital interest in defining the most appropriate accreditation criteria. The exact procedure to accomplish this task will evolve as the third-party accreditation process to design the system progresses. Certainly, the opportunity for the laboratory community to provide input into the process of defining the accreditation criteria is a minimal first step. Enlightened self-interest, if nothing else, on the part of all three groups — the third-party accrediting organizations, the requesting organizations and the laboratory community — will provide the impetus for the success of this interactive process. The role of each participant is to make suggestions concerning the appropriateness of the accreditation criteria. Under no circumstances can an accreditation or certification program be drafted without participation from all parties. It must be noted that the data purchasers have the final decision regarding accreditation. Many accreditation and certification programs can be drafted. However, final accountability resides with the acceptance and use of the system(s) by data purchasers — that is the agencies and other data purchasers that will approve of the accreditations and certifications by using them.

RESPONSIBILITY FOR SYSTEM LOGISTICS

Admittedly, the logistics and operational parameters associated with the proposed third-party accreditation process are formidable, but certainly not impossible. The formalized evaluation process requires a considerable effort in scheduling and coordination. Such a system exists at least in principle, in the Superfund Contract Laboratory Program

(CLP). This system employs entities to manage the sample flow to the laboratories and an ongoing system of performance evaluation and monitoring.

While the CLP and other systems have several excellent attributes, they all suffer from shortcomings of any fragmented operation. As a consequence, it might be most appropriate to define a generic approach to operation and logistics and leave the details of logistics for consideration by the requesting organization and the third-party accreditation organization. The best features of all existing systems could best be included using that approach.

In any case, the following general areas must be considered:

1. Application of the laboratory to be evaluated.
2. Approval of the requesting organization for the laboratory to be evaluated for baseline or special accreditation.
3. Scheduling of the evaluation process by the third-party accreditation organization and notification of laboratories to be evaluated.
4. Completion of the evaluation process and determination of the accrediting body that requirements were/were not met.
5. Forwarding of the results of the evaluation process to the requesting organization and/or the laboratories.
6. Acceptance or rejection of the "special" part of the accreditation recommendation by the requesting organization.
7. Notification of approval of results of the requesting organizations acceptance or rejection of the "special" requirements to the laboratories.
8. Follow-up with laboratories by third party to delineate problems.
9. Suggestions by the third-party for problem resolution to both the requesting organization and the laboratories, if desired by the requesting organization.
10. Periodic re-evaluation of the laboratory performance by one or more third-parties as determined by the third-party accreditation body and the certification and requesting organizations. Such re-evaluation includes site visits and monitoring of on-going performance by performance evaluation samples and data assessment.
11. Forwarding the required on-going monitoring information to the third-party accreditation and certification body.

12. Maintenance of a data base by the third party to assist in the evaluation of on-going laboratory performance and to provide the requesting organizations with trend analysis data.

Obviously, numerous areas listed above might be confidential to one or more of the parties and such data must be maintained in a secure manner. Some of the data should be available on-line to provide up-to-date information about all the accredited laboratories. This subject is discussed in greater detail in the following section.

Other parameters and operational logistics will need to be defined and the appropriate details provided. However, the logistical attributes presented above are critical to the successful operation of the proposed accreditation process. In addition, the requesting organizations and the third-party accreditation and certification organization(s) must interface closely and on a frequent basis. While the proposed accreditation and certification process will greatly reduce duplicate effort and increase the efficiency and effectiveness of laboratory accreditation, the requesting organization can not abdicate their responsibility or accountability. Consequently, the requesting organization must provide appropriate resources, to ensure the success of this interface.

DISTRIBUTION OF ACCREDITATION AND CERTIFICATION INFORMATION

In the past, one large accreditation-type program instructed anyone interested in the results of their program to ask the individual laboratories for information about their status, results of performance evaluation samples, on-going data review assessment and other information.[13] This was done to protect the laboratories and also to protect the government from harming the business of the laboratories that were not in good standing in the particular system. This approach was necessary because all information gathered about the laboratories was gathered as a result of necessary contractual oversight. As such, it was considered inappropriate for the government agency to make this information public.

This particular way of communicating certification-like information has caused the laboratory industry many problems. Since this government program conducts a rigorous program that utilizes performance evaluation samples, many entities, including other government agencies use the system as an accreditation program. In purchasing data from only laboratories that are a part of this system, the purchasers are using this special government program as a de facto certification system. At face value, this might not seem to be undesirable. However, because only laboratories that are awarded contracts are assessed by the contract oversight procedures, any other laboratories are specifically excluded from this system. Non-membership in this program does not mean a laboratory is not capable and able to meet the de facto certification criteria; it means only that the laboratory is excluded from evaluation under this program because it does not hold a contract.

The de facto certification system is easily recognized as the CLP. The CLP de facto certification system would not necessarily be a problem if there were an accreditation program that existed that is as least as rigorous as the CLP. All interested laboratories could then apply for certification and be assessed by the certification and accreditation authority(s). However, a suitably rigorous program does not currently exist. The effect of this is that laboratories bid low prices on CLP contracts in order to become "de facto certified." They want status in the CLP, because it enables them to market to other purchasers of environmental data that require de facto certification in their data purchase specifications.

One important lesson is learned from the CLP's methods of communicating information about laboratories. This is that the national program should communicate as much data as is necessary to provide ability and capability information about assessed laboratories.

A rigorous national accreditation program could alleviate the problems caused by de facto certification and could also allow on-line access to accreditation information by people that need to purchase data. For a set fee, which would offset some of the expense of database management, some access to both baseline inspection of capability "checklist" and initial performance evaluation sample analysis in addition to on-going monitoring status (ability) information could be provided to requestors.

The laboratories could also benefit from purchaser's access to long-term information. If a laboratory maintained consistent good status in the program, the effect of one "questionable" performance evaluation sample result could be viewed in the proper perspective by prospective clients. The results of consistent questionable standing in the organization could also be viewed by prospective clients. This would tend to improve the long-term quality of laboratory data, since obviously poor performing laboratories would need to improve quality to remain competitive. Since the laboratories are concerned about the possibility of being severely impacted by one bad performance evaluation result, the effects of a one-time problem could be alleviated by the proper display of past and current evaluation information.

APPEALS, ARBITRATIONS, AND LIABILITY

Recently, the question of appeals, arbitration, and liability associated with accreditation has surfaced as having the potential to become a real problem for an accreditation authority. Many of the laboratories that are a part of the CLP de facto certification program have undergone investigations that have impacted their business. It is sufficient to state that all third-party certification and accreditation programs need to address potential liability.[14] The need for appeals, arbitrations, and other such procedures must be worked out in the process of defining the certification and accreditation program and must be agreed to by all parties. The well established third-party programs are informed about these issues and can be called upon to assist in determining the best ways to handle them.

THE IMPORTANCE OF EDUCATION

In the past, accreditation programs have proliferated and existed in an environment in which each of the parties — the laboratories, the accreditation bodies, and the purchasers of data were more or less uninformed about the needs of all the other parties. For any new system to be successful, all organizations using the system or assessed by the system must be educated about the benefits and limitations of the system and the evaluation information that is provided by the accrediting and certifying authority(s). All parties must be fully informed about the accreditation and certification processes, the meaning of the tiers and how to interpret the information that is provided by the accreditation and certification authority(s).

Purchasers of laboratory data must be able to assess whether specific evaluation results are germane to laboratories' ability to produce quality data for a purchaser's specific needs. Data purchasers must use accreditation and certification as tools to facilitate cooperative efforts with laboratories to produce exactly the data that is needed. Accreditation and certification do not, and will never, assure data purchasers of quality data if the purchasers do not delineate their requirements to the laboratory effectively. Laboratories cannot produce quality results unless they are informed of the data user's needs. When data are ordered, there must be an active interplay between the data users and suppliers to assure that the correct data are obtained.

COST AND BENEFITS

As we have previously stated — and as has been demonstrated on numerous occasions — the unacceptable number of current systems of laboratory accreditation results in a great duplication of effort with only limited acceptance. Such a state of chaos results in greatly inflated cost of laboratory QA efforts, in terms of both time and money.[15]

By implementing a national program of third-party laboratory accreditation, all parties — purchasers of analytical data, the laboratory community, and the public — benefit. The purchasers of analytical data benefit from reduced cost because a widely accepted laboratory accreditation program will reduce overhead expenses in the laboratories that are passed on to the customers. The laboratory community benefits from reduced cost associated with accreditation and because their expertise, as demonstrated by the accreditation process will be widely recognized and accepted. The public will benefit most by the assurance that analytical data for any purpose — product specifications, clinical data, and so forth — will be of a well defined quality. It can therefore be assessed to determine if it is sufficient for its intended use. We truly have the opportunity for a win/win endeavor.

Now is also the time for this country to recognize the international benefits that will be achieved by national consistency.[16] The creation of the 1992 Single European Market[17] — the world's largest trading block — is quickly becoming a reality. In order to ensure the competitiveness of U.S. companies, national quality systems must be implemented. Further, these national systems must conform to the international requirements, if they are to be accepted in the international marketplace.

ADDITIONAL HURDLES TO IMPLEMENTATION

Government Contracting Requirements

Government purchasing requirements are a formidable hurdle to jump — formidable but certainly not impossible. Privatization has been championed for the past several years by many government agencies.[18,19] This move to privatize many service type functions that have previously been the purview of the government departments or agencies is a laudable effort. It has also resulted in the growth of a rather large segment of the commercial community with a great deal of business acumen regarding dealing with government purchasing requirements. Assuming the acceptance of the independent third-party accreditation approach, the government purchasing requirements will have to allow an independent third party assessment of baseline capability and product inspections to ensure the success of this approach. In addition, the political support of such an approach should be extremely broad if it is to succeed.

The Department of Defense (DOD) and their contractor base[20] have been struggling with the issues surrounding quality improvements, increased productivity, and scrap, rework and repair problems for several years. The Washington Bottom Line conferences that began in 1982 made it clear to industry that supplier's failure to improve quality would result in severe negative repercussions including loss of business.

In order to improve quality and productivity, The Ten Point Excellence Program was developed by DOD and the defense industry to outline the new means that customers and contractors will use to enhance product and service quality. This program opened an entirely new way of thinking about quality for the DOD and its contractors. The ten points of the program are

1. Design/build quality in
2. Award contracts based on quality
3. Streamline contract requirements
4. Modernize factories
5. Provide employee incentives for reducing scrap, rework, and repair
6. Increase and improve training and communication in the quality discipline
7. Implement guarantees

8. Enhance the quality assurance work force
9. Overhaul the government quality documentation (Handbook 50)
10. Tighten quality surveillance

These ten requirements bear a remarkable similarity to the Deming 14 Points for total quality management. Even though the process appears to be a step in the right direction to allow quality improvements, the reality of the implementation of the ten points is not so positive. Even though DOD has taken a proactive quality stand, and states that specifications should be reduced or tailored, the reality is that externally imposed management procedures are growing and contract specifications are becoming increasingly cumbersome. Overcoming the long-standing reluctance to reduce the multitude of contract specifications and streamline procedures will be difficult. Increased emphasis on design and production reviews that were intended to assure that quality is designed in are adding substantial new cost to contractor operations. One ramification of the increased number of audits is the increase in the number of interpretations of the contract requirements that are made by auditors. These conflicting interpretations often result in extra program cost with no added product quality. The American Society of Aerospace Engineers is seeking to convert some routine audit operations to a national contractor certification board. This measure is sought to help reduce cost and increase quality more than the existing set of redundant audits by both contractor and government.

This example about the DOD's struggle to implement quality improvement is presented to show that even with broad based political support, the implementation of total quality is an uphill struggle. The elements of total quality must be understood by all parties before it can be implemented correctly. DOD has taken appropriate first steps to implement changes, and most of the roadblocks to quality improvements have been removed. Still the goal of total departure from past practices is a slow and arduous process that can only be accomplished when government purchasers and contractor suppliers embrace, understand, and implement total quality management. The experiences of DOD should be used to design an accreditation and certification system that does not suffer from the same problems as this DOD effort.

Government's Use of Third Parties

One issue that must be resolved before third parties can be requested to assist in this effort is the question, "Is it feasible to use third–party accreditation and certification bodies to assist in developing standards for accreditation and certification?" This issue has been raised in the past and the question has already been answered by others. It most certainly is feasible. This process has been used in other areas. The American Society for Testing Materials (ASTM)[21] has assisted the EPA Office of Solid Waste in developing quality assurance specifications in the past. The standard practice for generation of environmental data related to waste management activities is being developed in cooperation with the ASTM D34 committee.

EPA has also elected to use the third–party route in development of standards and to test and certify additives for drinking water. This is one example of how the third–party process can work for production of standards for certification and accreditation of testing laboratories and the subsequent testing and certification of the laboratories and their products in accordance with the standards.

Funding of Third-Party Efforts

Who pays for the cost of the third-party standards development process? In the case of NSF programs,[22] the fees are set to provide only full cost recovery for the development process. The cost is generally pro-rated to manufacturers that participate in the standards development process. Companies that later apply for listing under the developed standard pay the fee prior to listing. This collection of fees provides funding to review and revise the standard at intervals that do not exceed 5 years.

In the future, the cost of third-party program development could need to be reassessed to assure rapid development of the standard criteria and appropriate review and update of requirements. As is true with most projects, the faster the process for development, the more resources are expended. In order to speed the development, a larger share of initial expenses might need to be provided by the government. The lowered cost for overall, long-term program to the taxpayer would justify this initial expense.

Government's Lack of Personnel to Fully Address the Issue

Laboratories, industries that purchase data, and government agencies can work together to effectively draft and implement a solution. However, the government cannot bear all the burden. Clearly, national accreditation is seen as a high priority for environmental laboratories. However, the government might not view this problem in such a light. There are not enough people to address all critical problems in EPA effectively. Therefore, this problem might not be addressed soon. There are simply not enough resources to address issues that are not perceived to be the most critical problems on the list of EPA's priorities.

Because of this lack of enough people to address all issues, the government can use the third-party approach to improve the situation for both the government and industry. If government will request assistance from a credible third party, the lack of government personnel to move this issue from inaction to action will not be a problem.

REFERENCES

1. "Availability, Adequacy, and Comparability of Testing Procedures for the Analysis of Pollutants Established Under Section 304(h) of the Federal Water Pollution Control Act, Report to the Committee on Public Works and Transportation of the House of Representatives, EPA/600/9-87/030, p. 1.9 (1988).
2. "Proceedings of the USATHAMA Quality Assurance In Environmental Measurements Meeting — Las Vegas, NV," prepared by Science and technology Corporation, Hampton, VA. (May 1989).
3. "Accreditation Committee Update," *Environmental Testing ADVOCATE*. II(1): 1,4 (1990).
4. Warshaw, Stanley I., "International Laboratory Accreditation Conference," *ASTM Standardization News* 14(1): 42–44 (1986).
5. Hess, Earl, H., "Laboratory Accreditation," *ASTM Standardization News* 14(1): 31–34 (1986).
6. "Accreditation Committee Update," *Environmental Testing ADVOCATE*. II(1): 1,4 (1990).
7. Fisher, Steven A., "Certification and Accreditation," *Environmental Lab* 1(4): 38–39 (1989).
8. Hess, Earl, H., "Certification and Accreditation," *Environmental Lab* 1(4): 41–42 (1989).
9. Blacker, Stanley M., "Data Quality and the Environment," *Quality*, pp. 38-40 (April 1990).

10. Carlberg, K. and A. Babyak, "Meeting Presentation at the International Association of Environmental Testing Laboratories Accreditation Subcommittee Meeting," Wilmington, DE, February 6, 1989.

11. "Comparison of the New York Environmental Laboratory Approval Program (ELAP) With Other Quality Assurance Programs," Report to the New York State Division of the Budget prepared by Dynamac Corporation (1988).

12. "Availability, Adequacy, and Comparability of Testing Procedures for the Analysis of Pollutants Established Under Section 304(h) of the Federal Water Pollution Control Act, Report to the Committee on Public Works and Transportation of the House of Representatives, EPA/600/9-87/030, p. 1.4 (1988).

13. Miller, M.S. and M. Stutz. "Quality Assurance/Quality Control Task Force Update" in *Proceedings of the United States Army Toxic and Hazardous Materials Agency (USATHAMA) Quality Assurance In Environmental Measurements Meeting,*" prepared by Oak Ridge Gaseous Diffusion Plant by Martin Marietta Energy Systems, Inc., Oak Ridge, TN., p. 91 (1988).

14. Jacobs, Richard M., "Products Liability: A Technical and Ethical Challenge," *Quality Progress*, pp. 27–29 (December 88).

15. Farrell, John, "Do Multiple Certification and Accreditation Programs Enhance the Quality of Environmental Data Generated by Commercial Laboratories — An Alternate Approach" presented at the USATHAMA Quality Assurance In Environmental Measurements Meeting, Las Vegas, NV (May 1989).

16. Yaus, Leo P., "The Global Search for Quality", *Quality Progress* pp. 51–53 (August 1988).

17. Boehling, Walter H., Europe 1992: Its Effect on International Standards," *Quality Progress*, 23(6): 29–32 (1990).

18. "EMSL Privatizes Two Functions," *Environmental Testing Advocate* II(1): 2,4 (1990).

19. "Privatization — Toward More Effective Government, Report of the President's Commission on Privatization (March 1988).

20. Talley, Dorsey J., "New Challenges for Quality Professionals in Defense Industries" *Quality Progress*, pp.41–43 (December l988).

21. Tait, Reid, "Memorandum to ASTM D34.02 Members - Subject: Ballot for Emergency Standard DRAFT ASTM Standard," (February 16, 1989).

22. McClelland, Nina I., "The National Sanitation Foundation's (NSF's) Third-Party Programs to Assist States," presented to the Association of State Drinking Water Administrators (ASDWA), Tampa FL, p. 4 (February 1989).

Chapter 9

PERFORMANCE EVALUATION MATERIALS — A VITAL ELEMENT OF THE SYSTEM

INTRODUCTION

The use of quality control materials, performance evaluation (PE) materials (blind samples) and reference materials (RM) has been an accepted practice for determining the ability of testing laboratories to perform analyses correctly for as long as QA and QC has been practiced. As has been pointed out previously, the use of these materials to evaluate the performance of laboratories in a national accreditation and certification program is a critical requirement. The results of analysis of the performance evaluation samples lends credibility and acceptance to laboratory accreditation programs and provides results that serve as an integral part of data review and certification.

DEFINITION OF TERMS

Before we discuss the positive attributes associated with performance evaluation samples, it is beneficial to review and define the appropriate terms. Many programs and offices within EPA and other environmental programs in other federal and state agencies have different names for each of the materials used to assess laboratory performance. Rather than use the names that are specific to EPA and other programs, the definitions for these materials have been taken from a widely accepted reference used throughout analytical chemistry laboratories, Quality

177

Assurance of Chemical Measurements. In this book,[1] John Taylor provides the following definitions:

> Blind sample — A sample submitted for analysis whose composition is known to the submitter but unknown to the analyst. A blind sample is one way to test proficiency of a measurement process. (Note from the authors — these materials are referred to as Performance Evaluation (PE) samples or Performance Evaluation Materials (PEMs) in the EPA CLP. The CLP utilizes mostly single blind samples, but endeavors to use double blind samples. The use of double blind samples is encouraged. In other chapters of this book, the term blind sample can be substituted for PE sample or PEM).
>
> Certified reference material (CRM) — A reference material, one or more of whose property values are certified by a technically valid procedure, accompanied by or traceable to a certificate or other documentation which is issued by a certifying body.
>
> Control sample — A material of known composition that is analyzed concurrently with test samples to evaluate a measurement process.
>
> Double blind — A sample known by the submitter but submitted to an analyst in such a way that neither its composition nor its identification as a check sample are known to the latter.
>
> Duplicate sample — a second sample randomly selected from a population of interest to assist in the evaluation of sample variance. (See also Split sample.)
>
> Performance audit — A process to evaluate the proficiency of an analyst or laboratory by evaluation of the results obtained on known test materials.
>
> Primary standard — A substance or artifact, the value of which can be accepted (within specific limits) without question when used to establish the value of the same or related property of another material. Note that the primary standard for one user may have been the secondary standard of another.
>
> Reference material (RM) — A material or substance, one or more properties of which are sufficiently well established to be used for the calibration of an apparatus, the assessment of a measurement method, or for the assignment of values to materials.
>
> Secondary standard — A standard whose value is based upon comparison with some primary standard. Note that a secondary standard, once its value is established, can become a primary standard for some other user.
>
> Split sample — A replicate portion or subsample of a total sample obtained in such a manner that it is not believed to differ significantly from other portions of the same sample.
>
> Standard reference material (SRM) — A reference material distributed and certified by the National Bureau of Standards.

Each of the materials defined above plays a role in the total quality approach to ensuring the generation of data of acceptable quality.

However, for the present discussion, these definitions are presented in an effort to alleviate the widespread confusion associated with the use of such materials and to allow the terms to be used without further explanation in this chapter.

CURRENT USE OF BLIND SAMPLES IN LABORATORY EVALUATION PROGRAMS

The EPA's CLP currently utilizes quarterly blind samples that are called Performance Evaluation (PE) samples. The EPA Office of Solid Waste is in the process of devising a means to more fully utilize blind samples in their data collection quality assurance procedures.[2] The United States Army Toxic and Hazardous Materials Agency (USATHAMA) operates a large contract laboratory program. While USATHAMA has a QA program similar to the CLP, it does not conduct a periodic blind sample program. However, this Agency recognizes the merits of such a program and is actively pursuing adding this requirement to the USATHAMA QA program.[3]

How do blind samples contribute to the overall success of a national third-party accreditation and certification program? Blind samples provide the benchmark for determining the performance of testing laboratories and are a cornerstone in the process of data reviews and certification. These materials must be used during the evaluation of the ability of laboratories to perform. The results of analyses of the blind samples provide an objective assessment of ability in addition to providing the foundation for problem resolution and corrective action if problems are found in the laboratory's results on the blind samples.

As we have emphatically stated in several previous discussions, the success of a national accreditation and certification program is critically dependent on credibility and acceptance. We will restate the obvious at this juncture — the incorporation of blind samples in an accreditation and certification program virtually ensures its success. The use of realistic blind samples provides a means for defining the ability of laboratories to perform. The judicious choice of blind samples allows the evaluation organization to both assess the laboratory's performance and to objectively define problem areas.

The use of PE samples by the users of an accreditation and certification system provides oversight of the effectiveness of the accreditation and certification program. These materials, therefore, make it possible to determine whether a certification and accreditation system is working and is relevant to the data purchaser's needs.

REAL MATRIX SAMPLES VERSUS SYNTHETIC SAMPLES — ARE BOTH TYPES VALID?

Blind samples can be either fortified matrices or "real world" samples. "Real world" blind samples consist of a matrix and analytes that have combined with each other in nature. In other words, the constituents are not spiked onto the matrix in a laboratory, but are "naturally" incorporated into the matrix of interest. Fortified matrix samples, also called "spiked" or "synthetic" blind samples consist of a matrix and analytes that have been combined together in a laboratory-type production process, rather than in a natural setting. These materials might consist of "real world" blind material that have been fortified by spiking with additional analytes or to increase the levels of naturally incorporated constituents. Both types are useful and provide information that is critical to a successful accreditation and certification program. The materials provide different information about laboratory performance. The materials that most closely mimic the types of samples that are analyzed are more useful to determine laboratories' ability to perform on specific types of samples. Care must be used to select most appropriate PE samples and interpret their results. However, all materials can be used to provide an assessment of laboratory ability.

There has been a long-standing debate among users of the both synthetic and real matrix materials about which type of material provides the best information about the analytical test. It is obvious that, in order to assess the actual matrix of the samples being tested, it is important to match the quality control samples as well as possible to the real matrix conditions of the actual samples. This is critical if the quality control samples are the only quality control samples to provide information about the test. However, in most cases, there are several other measures of performance that can provide supplementary information about the analysis. If one can place a "perfect" blind sample — that is one that has a matrix and analyte composition that is exactly like the samples — in each set of samples, then most usual QC samples might not be as important as they currently are. That means that the quality control information that matrix spikes, matrix spike duplicates, laboratory control samples, and internal standard spikes provide would only confirm what the PE sample results provided. However, since in most cases, exact sample, analyte, and matrix matches are not made, additional QC samples provide the information that is needed to correctly interpret the blind sample results and how their analyses relates to the interpretation of the real sample results. The better the blind samples mimic the samples in the batch, the less critical the interpretation and utilization of other QC information becomes.

The key to successful use of both types of materials is to determine how closely the matrix materials match the samples that are being analyzed and then determine how much additional QC is needed and how to effectively use the QC information. In many cases, QC samples and procedures are required by regulatory methods. In such cases, the chemist can still determine how to interpret the results of the analysis based on all QC, including the use of PE samples. In addition, the chemist can determine whether additional QC is needed to effectively characterize precision and bias when regulatory methods are used for production of data for special projects. Careful analysis of QC requirements sometimes might indicate that more or less QC is needed in specific laboratories and for specific samples if information about precision, bias, and quantitation limits are to be correctly characterized to facilitate data use.

In the future, the use of PE samples will be more prevalent, as the EPA determines how and when these materials are best used in environmental analyses.[4] This will streamline and add significantly to internal laboratory QA and QC procedures. In addition, the judicious use of PE samples will aid certification and accreditation authorities in their assessments of laboratory ability.

INTERNAL VERSUS EXTERNAL QC SAMPLES

In the past, many skeptics that discount the use of PE samples have stated that there is no way that analyte and matrix combinations that match samples sufficiently well can be made. Therefore, they have discounted the importance of such samples in assessment of laboratory performance and have relied solely on internal QC sample analyses to provide quality information. Some laboratories analyze internal quality control samples in addition to the usual matrix spike/matrix spike duplicates and internal standard spikes in their analytical sequences.

These analyses are intended to provide internal laboratory QC information. They do not provide an independent external assessment of laboratory quality. The intent of laboratory accreditation and data certification is to provide data purchasers with independent assessments of quality. Internal quality assurance techniques, including use of PE samples are necessary to control day-to-day quality. However, outside, independent checks are necessary to provide un-biased verification of laboratory ability that is relative to outside, not internal, standards.

It is true that not all matrix and analyte combinations of blind PE samples are available, but the use of available materials provides information about the laboratories' ability to perform on some matrices and analytes. This information can be used to project possible ability on similar matrices and analyte combinations. The lack of availability of all materials is not a reason to exclude these materials from a laboratory evaluation program. Several general types of matrices containing specifically chosen chemicals can provide information about laboratories' ability to analyze many other chemicals. For instance, one does not need to include all chemicals of interest in a matrix to ascertain that a laboratory can correctly analyze for all the analytes. Instead, selected chemicals can be added to the matrix because they are known to be easy to identify, difficult to extract, easily degraded if the analysis is not performed correctly, or for a multitude of other reasons. The laboratory's performance on these analytes can be evaluated and predictions about whether other analytes would be correctly analyzed by the laboratory based on the laboratory's success on the PE sample can be made.

THE USE OF PERFORMANCE EVALUATION MATERIALS BY INDUSTRY

Since the best evaluation of performance occurs when the laboratories are challenged with blind samples that are as close to the samples that are submitted for analysis as possible, it is ideal if data purchasers submit well-characterized samples of their industrial or site-specific wastes to the laboratories as blind samples. The sample results can be assessed based on past laboratory evaluations of the same sample and the data purchasers can determine if the laboratories are producing data that is reasonable based on past analyses of the blind samples. The data purchasers and the laboratories can work together, if necessary, to determine why and where any problems occurred in the analyses to improve future matrix-specific analyses.

THE OPTIMAL SOLUTION — USE OF PERFORMANCE EVALUATION MATERIALS BY BOTH ACCREDITATION PROGRAMS AND DATA PURCHASERS

Note that the above explanations of the use of blind PE samples indicate that generic matrices and analytes can be used to provide general accreditation and certification program information about the ability of laboratories. Specific matrix and analyte information can be obtained by data purchasers if they include PE samples in their sample batches that are sent to laboratories. The need for inclusion and mass production of industry-specific materials is not necessary for accreditation and certification purposes, but is certainly an excellent practice for data purchasers. As was stated previously in this chapter, data purchaser's use of PE samples provides an assessment of the relevance and effectiveness of the generic baseline accreditation and certification program.

USE OF PERFORMANCE EVALUATION SAMPLES VERSUS SPLIT SAMPLES

At this point, it is important to stress the difference between using well-characterized "real-matrix" samples from the practice of sending split samples to different laboratories. In the past, some data purchasers split a single sample into two or more samples that were then analyzed by different laboratories. The sample results were compared and it was determined if the samples were correctly analyzed by one or all laboratories. The difference between sending out true blind PE samples and split samples is that, in the case of the split samples, one does not know which laboratories' results are correct if the results from the laboratories differ. In fact, if the results do not differ, one does not know if the results are truly accurate values, because the "true" value is not well characterized through interlaboratory studies of the material. If true blind samples are sent to several laboratories, one knows the "true" values and can compare the results to the "true" value in order to provide a valid assessment of the differing laboratory results.

USE OF PERFORMANCE EVALUATION MATERIALS TO ASSIST IN DATA REVIEW AND ASSESSMENT

In addition to providing data on the laboratories' ability to perform the testing functions being monitored, the inclusion of blind samples with each lot of samples analyzed provides a benchmark for data review. Consider for a moment the monumental task of reviewing data generated by programs such as the CLP and USATHAMA. This effort currently requires time and expertise of many data reviewers. If, however, an objective benchmark — a blind sample — were available for initial review, a determination about the quality of the batch level of the data becomes possible. Rather than evaluate each attribute or requirement as it is currently performed in data review procedures, an initial determination based on the results of the blind sample would provide almost instant data useability information. In addition, data from the analysis of specific analytes provides information that assists in determining where problems occurred and if any data resulting from a flawed analysis could be useful. The use of blind samples in this manner not only protects the purchaser of data but also protects the laboratory. Rather than rejecting all data, it becomes possible to use some data from flawed analyses because it might be possible to determine what data are affected by the problems that occurred in the laboratory analysis and what data are not affected by the problems. The laboratory can therefore be compensated for the data that is useful, and can correct the problems on future analyses.

In the future, many of the time-consuming steps involved in data review will be computerized. However, this computerization will not totally replace the need for a chemist's evaluation of data. Both computerized data review and assessment of data by a chemist will be enhanced by the use of PE samples. In fact, this data could provide the needed information for an expert system approach to determining what results can be used and which results cannot be used for specific purposes. The data from PE samples could also allow large quantities of data to be screened to determine which sets of data need a thorough evaluation, which sets of data are totally useless with no further evaluation needed, and identify areas for further evaluation based on problems identified by the computerized assessment. Any screening and identification of problems by a computer system will conserve valuable resources and reduce the time required for data evaluation.[5]

PERFORMANCE EVALUATION MATERIALS SUPPLIERS

Using blind samples as described in this chapter is beneficial to both data purchasers and users. However, this use makes it necessary for the suppliers of materials to maintain the production of blind samples in a manner that ensures integrity — not only of the blind samples, but also of the supplier(s)' organization. The SRMs that can be obtained from NIST are examples of such materials. Unfortunately, NIST is not able to supply all types and matrices of blind samples in a reasonable time frame and within a cost structure that allows for routine use of the materials.

A limited number of commercial operations currently manufacture reference materials for QC use and PE samples for a selected set of analytes. The manufacture and sale of the materials has been limited in the past, partially because the materials are not required for analyses in conjunction with regulatory analyses. The manufacture has also been limited partially because no standard set of requirements for evaluation of the quality of the materials exists. Therefore, the manufacturers could not identify their products as meeting EPA requirements or otherwise indicate that the products were of a certain quality level. Both of these limiting factors might be remedied soon. The EPA is currently sponsoring the development of a third-party consensus product standard for the materials. The EPA's Office of Solid Waste (OSW) further plans to sponsor the development of a certification standard for the products in order to promote their manufacture and use in environmental analyses.[6]

The manufacture of these materials is relatively uncomplicated for many materials. The possible exception is the manufacture of real matrix volatile materials. The facilities and equipment requirements for the manufacture of PE samples is reasonably limited. The manufacture of industry-specific materials are the easiest type of material to formulate, because these materials often do not require "fortification" and in-house verification of the analyte levels. Industry-specific materials can be obtained from appropriate industrial sources, blended, analyzed, bottled and sealed and provided to clients. The "windows" of acceptance for the analytes of interest can be determined by limited interlaboratory testing. Concomitant with preparation of PE samples, a stability and homogeneity testing program is necessary to ensure the integrity of the materials. Following from this approach, PE samples can be devised for generic laboratory evaluations, problem resolution and corrective action procedures, method specific evaluations and data review. The options available are limited only by the imagination of the supplier and the acceptance of the purchaser.

COST OF PERFORMANCE EVALUATION SAMPLES

In the past, the cost of PE samples has dissuaded many programs and individual industrial data purchasers from utilizing the materials. However, a cost and benefits analysis of the use of the materials should be performed to assess just how cost effective this practice is. For instance, if use of the materials enables a program to evaluate laboratory performance with minimal on-site assessment, the cost of reduced on-site assessments can defray the cost of the samples. In addition, if the evaluation of PE sample data allows screening data for defects and provides as a less labor-intensive, data-package inspection, this also provides a substantial cost savings. In addition, if data can be better used and assessed because the problems that affect data and those that do not affect data can be correlated to analytes in the PE samples, data can be used to the maximum extent possible. The cost savings brought about through better use of data could be substantial. Finally, the use of PE samples might substitute as a more effective QC than other QC procedures. In the future, the use of the samples in lieu of other required QC might be allowed. This concept will be tested by EPA's Superfund Program by assessment of data from one of their contract methods.[7]

USE OF PERFORMANCE EVALUATION MATERIALS IN THE PROPOSED ACCREDITATION AND CERTIFICATION PROGRAM

Inclusion of blind PE samples in the third party accreditation and certification program is vital to the success and credibility of the program. The use of the materials provides tangible evidence of the performance or lack of performance of laboratories. It is not necessary for the third party to produce the required PE samples, only that they have access to suitable materials. Because the inclusion of PE samples is an integral part of a credible accreditation program, the widespread acceptance of these materials will be assured. The judicious use of PE samples will enhance a national accreditation and certification program and improve the performance of the testing laboratories involved in the program.

REFERENCES

1. Taylor, J. K., *Quality Assurance of Chemical Measurements* (Chelsea, MI: Lewis Publishers, Inc., 1987) pp. 245-253.
2. "Summary of the Reference (RM) Certification Group Meeting of Tuesday, May 8, 1990," Memorandum to Reference Materials Task Group, A2LA Accredited Environmental Laboratories, A2LA Environmental Assessors, A2LA Environmental Advisory Committee, Reference Material Suppliers, A2LA Board of Directors and Other Interested Parties, Memorandum distributed by the American Association for Laboratory Accreditation, Gaithersburg, MD (May 23, 1990).
3. Ryan, A., "Quality Assurance/Quality Control Task Force Update" in *Proceedings of the United States Army Toxic and Hazardous Materials Agency (USATHAMA) Quality Assurance In Environmental Measurements Meeting,"* prepared by Oak Ridge Gaseous Diffusion Plant by Martin Marietta Energy Systems, Inc., p. 25 (1988).
4. Gebhart, Judith E., "External Monitoring Through Use of Performance Evaluation Samples," paper presented at the Environmental Hazards Conference and Exposition, Seattle, WA (May 1990).
5. Flynn, May J., Carla R. Schumann, and Ramon A. Olivero, *Computer-Aided Data Review and Evaluation - CADRE - Release 1.01*, prepared by Lockheed Engineering and Sciences Company, Environmental Programs, Las Vegas, NV for the U.S. Environmental Protection Agency, Environmental Monitoring and Support Laboratory, Quality Assurance Research Branch, Las Vegas, NV (1990).
6. "Summary of the Reference (RM) Certification Group Meeting of Tuesday, May 8, 1990," Memorandum to Reference Materials Task Group, A2LA Accredited Environmental Laboratories, A2LA Environmental Assessors, A2LA Environmental Advisory Committee, Reference Material Suppliers, A2LA Board of Directors and Other Interested Parties, Memorandum distributed by the American Association for Laboratory Accreditation, Gaithersburg, MD (May 23, 1990).
7. Gebhart, Judith E., "External Monitoring Through Use of Performance Evaluation Samples," paper presented at the Environmental Hazards Conference and Exposition, Seattle, WA (May 1990).

Chapter 10

APPLICATION OF THIS SYSTEM TO NON-ENVIRONMENTAL LABORATORIES

COMPARISON OF THE MISSION OF NON-ENVIRONMENTAL TESTING LABORATORIES TO ENVIRONMENTAL TESTING LABORATORIES

The approach to laboratory accreditation described in previous chapters used the environmental testing laboratory as a model template. Due to the complexity of the analyses provided by the environmental laboratory, it functions as an ideal model for development of the proposed approach to laboratory accreditation. Moreover, the environmental testing laboratory embodies all facets of the laboratory community. Like many government laboratories, much of the analytical data produced by the environmental laboratory is regulation driven and as such could be an integral part of litigation proceedings. In a vein similar to industrial laboratories, the analytical data produced by the environmental laboratory is almost always intimately involved in business decisions potentially costing millions of dollars. Like the clinical laboratory, the analytical data produced by the environmental laboratory often has the potential to affect the health and well being of many individuals.

Recognizing the critical nature of analytical data and the similarities to a broad array of laboratory types inherent in the products of the environmental testing laboratory, it is prudent to suggest that the proposed laboratory accreditation process can be successfully applied to all laboratory types.

KEY ELEMENTS OF THE SYSTEM THAT APPLY TO ALL LABORATORY TYPES

All laboratories that produce data do so by following a process to produce the data. Therefore, similar questions about laboratory capability and ability are germane to all laboratory types. Two critical questions must be asked:

1. What is the capability of the laboratory to produce a quality product?
2. What is the ability of the laboratory to produce a quality product?

These questions can be addressed by the process that has been previously described for environmental laboratories in Chapter 7.

Briefly, the proposed third party approach to laboratory accreditation embodies the attributes required to evaluate any laboratory type. As a review, these attributes fall into the following categories:

Accreditation Criteria

I. Generic Capability Indicators
    A. Organization
    B. Personnel
    C. Facilities
    D. Equipment
    E. SOPs
    F. Reference Standards and Performance Evaluation Materials
    G. Chain-of-Custody
    H. Documentation and Records Retention
    I. Data Reporting Format and Content
    J. QA Plan

II. General Ability Indicators
    A. Proficiency Evaluation (PE) Sample(s)
    B. QA Audit Based on PE Sample Results
    C. Data Review

APPLICABILITY OF THE CONCEPT

It remains for the various organizations currently involved in laboratory accreditation to facilitate the utilization of the third–party approach. All the attributes of this process described previously for environmental laboratories are germane to the accreditation of all laboratory types. Any desired criteria required can surely be included in the specifications and consensus standards provided by the organization requesting the accreditation evaluation.

It is important to also point out that many testing laboratories are currently utilizing multiple systems for the purposes of ascertaining if the quality of data produced by the various types of testing laboratories are of sufficient quality. For many of the same reasons that environmental testing laboratory accreditation and certification systems are not currently successful, these other systems are also not totally successful. It is important to remember that all criteria used when conducting inspections to determine capability of the process must be correlated to the defect that will occur in the product if the process fails at that inspection point. Specified requirements must always be necessary and they must be sufficient to produce a product or service of the needed quality.

Chapter 11

THE IMPACT OF NATIONAL LABORATORY ACCREDITATION ON INTERNAL QUALITY ASSURANCE AND CONTROL

INTERNAL LABORATORY QUALITY ASSURANCE — RESEARCH AND DEVELOPMENT QA VERSUS "CONTRACT CHEMISTRY" QA

Every well-managed laboratory has a comprehensive internal quality assurance program. The quality assurance program can be formal and routine as is usually found in production laboratories, or flexible and informal as is usually found in research and development laboratories. Some aspects of the quality assurance programs used in research and development laboratories have their place in a production laboratory. Special projects frequently have the same requirements as a research project. Therefore, statistical quality control production requirements are not suitable. The person managing the project for the laboratory and the client that needs the data must determine the type and level of control that is optimum in such situations. Each type of program is useful for determining and documenting the quality of the data produced.[1]

It is useful to point out the positive influences of research and development type quality assurance practices. The level of quality achieved in development efforts is usually closely correlated to the skill and technical expertise of the individual chemists that are conducting the investigation and the skill and expertise of the peer review that the investigation receives. The quality of data needed is usually tied integrally to the questions to be answered with the data. Therefore the technical specifications of the data, such as precision, bias, sensitivity and other such requirements are based on real technical needs. The research is conducted to maximize the information content of the data. Quality control samples are included to define the uncertainty inherent in the data, in order to allow its use for making decisions.

Consider now the negative influences of production type quality assurance practices. The quality control data are gathered to define uncertainties in the data, but the uses of the data are not intrinsically tied to the quality specifications for which the data are purchased. Production laboratories provide data that meet generic quality specifications that are delineated in a contract or purchase agreement. This type of generic data is used to support many different decisions. Therefore, some of the quality control data might be used to assist in determining fitness of the data for use, while other QC data might be of little or no use to the project. The optimal QC, detection limits, sensitivity and other requirements are probably not obtained for each project; however the data quality is well defined, the limits for use are established, and the methods are well defined. These data are not produced by the same careful research approach that the research project data are produced. Rather, the chemical laboratory produces the results as an assembly-line process.

"Contract Chemistry" QA is not all negative. Several positive aspects far outweigh the negative aspects, if the laboratory does not relegate technical credibility of data to secondary importance after contractual requirements. The use of routine QC practices can optimize the quality of routine sample production. The laboratory can maintain statistical quality control over the process and optimize production at the same time that quality is controlled. Any deviations from QC requirements can be quickly discovered and corrected.

Another positive aspect for the purchasers of data that occurs with "contract chemistry" QA is that the methods and QA procedures employed in production processes are usually mandated. Therefore, the data produced from use of the methods is not challenged because the methodology is recognized as acceptable. One critical positive aspect of "contract chemistry" is the establishment of well understood and standardized data reporting requirements, raw data documentation and retention requirements, and chain-of-custody requirements for the samples and the resultant data. In order for data to be used for any purpose, the link between the sample and the resulting data must be established and maintained. In order for the data to be accepted for litigation purposes, the quality of the documentation supporting this link between sample and data determines whether the data are admissable as evidence and what weight they carry.[2] "Contract chemistry" procedures for establishing chain-of-custody and for reporting data are concise, accepted, and do not require the expert testimony of individuals that produced the documentation to clarify the documentation. Even though the actual procedures that are required by contracts might be disputed as inefficient or costly, the impact of "contract chemistry" on making sample tracking and documentation a routine practice is a positive aspect of this type of QA.

Clearly, the benefits of "contract chemistry" QA and QC are not enjoyed by research and development QA and QC. The documentation requirements are not routine. Documentation usually is comprised of detailed notes in a logbook. Chain-of-custody, that establishes a link between the samples and the results is required, but might not be as formalized as the "contract chemistry" samples. The QC requirements and the methodology are not routine; therefore, use of research and development type methods might require expert testimony to establish credibility of the methodology. Clearly these drawbacks to the research and development type of QA deter the use of this approach if the "contract chemistry" approach to data acquisition and QA will suffice.

The rest of this chapter describes the rudiments of formal internal quality assurance programs. This is because formal programs will be impacted most by an external national quality assurance program. It is important to realize that large-scale accreditation and certification programs cannot consider every type of methodology that is performed in each laboratory. Therefore, the research and development type quality assurance program is usually only assessed internally and by conscientious purchasers of data. However, it is important to point out that if a laboratory's formal quality assurance program is assessed by an accreditation authority and found to be deficient, an inference can and probably will be made by many laboratory customers that the laboratory's research and development quality assurance program is also deficient. In many cases, this inference may not be valid. However, this inference will probably still be made. A national baseline quality assurance program provides information that determines capability, and it can also be used to infer that a laboratory is not capable of maintaining a quality assurance program — either formal or for research and development projects.

ELEMENTS OF A FORMAL INTERNAL QUALITY ASSURANCE PROGRAM

A formal internal quality assurance program evaluates all critical elements that were described in Chapter 5, for the Contract Laboratory Program (CLP). In addition, some critical elements are evaluated in internal QA systems that the CLP on-site visits do not address. Two such elements are internal document storage procedures, and traceability and equivalency of standard reference materials, control samples, and other reference materials that are utilized to establish and maintain control of accuracy and precision of analyses. In addition, the chain-of-custody procedures and methods required by the CLP are routine and specific for CLP work. Therefore, the analytical methods and custody procedures, likewise, would not be assessed as critically by the CLP as an internal QA system would require.

The reason that the CLP does not assess the data storage and retrieval area in the detail that is required in internal QA assessments is that the CLP would not be significantly impacted by a lack of adherence to good practices in this area. This is because CLP data and documents are not stored at the laboratory. Rather, they are removed in their entirety from the laboratory and are stored by EPA.

Likewise, the CLP would not be significantly impacted if a laboratory did not control the standard reference materials, performance evaluation samples, and other such materials in its non-CLP operations. The CLP does not assess the traceability and quality of reference standards and control samples that are used by the CLP laboratories for CLP analyses because, in the past, these materials have been supplied by EPA. Since the traceability, equivalence, and quality of these materials was well known by EPA, the assessment of procedures to establish their quality was unnecessary. Therefore, only a limited assessment of laboratory standards handling and storage procedures is accomplished by the CLP on-site assessors.

The internal laboratory data evaluation function that determines data conformance with contract specifications has only recently been included in on-site evaluations. Clearly, this function should receive critical review by the internal quality assurance program because it is the final internal review that can catch defects before the client receives the product. While, in the past, the CLP has provided only cursory examination of this system, this area is now considered an important element of internal QA systems to carefully assess during future on-site CLP assessments.[3]

INTERNAL QUALITY ASSURANCE — WHY IT IS NEEDED?

Current Importance of Internal QA

Internal quality assurance currently is important in laboratory operations. It can identify problems with data so that defective data are not supplied to clients, therefore increasing the credibility of the laboratory with its customer base. It can also identify key problem areas in the laboratory so that improvements can be made to the system to optimize quality and productivity. Internal quality assurance programs can gather data needed to establish control limits. QC procedures for analyses can limit the number of repeat analyses to the minimum, therefore eliminating many costly reruns. In the final analysis, internal quality assurance provides information necessary to support a laboratory's claims about the quality of data resulting from the analyses it performs. Internal quality assurance data establishes a record of the laboratory's adherence to the laboratory's quality assurance program requirements, deviations from such requirements, and the effect on the data of such deviations. Internal quality assurance records can establish a laboratory's credibility if the results from a laboratory are challenged.

Future Importance of Internal QA

The last sentence in the last section is restated, "Internal quality assurance records can establish a laboratory's credibility if the results from a laboratory are challenged." If a national external quality assurance assessment program — that is, a national accreditation and certification program — is developed, many of the current programs should disappear. Currently, if a laboratory loses accreditation from one state, agency, or third party, it does not usually severely impact the business of the laboratory. Consider, however the impact that loss of accreditation from a credible, national third-party program could have. The loss of business could be significant, especially if the program has replaced many of the other assessment programs.

A laboratory must maintain adequate documentation of internal quality assurance practices to provide evidence to an accreditation and certification authority that a mistake has occurred if the authority finds a problem that the laboratory refutes. In addition, if the authority identifies a legitimate problem with the laboratory or the laboratory's data, the laboratory must be able to quickly identify the cause of a problem and correct the problem. The identification and correction of the problem will be faster and easier to accomplish if the laboratory maintains an internal quality assurance program. Past internal QA data that documents results of internal control samples, traceability and equivalency of standards information, instrument maintenance records and many other sources of data can be used to assist laboratory management in quickly identifying possible causes for the problem. Corrective action can be quickly completed and supporting documentation can be provided to the accreditation and certification authority.

A good internal quality assurance program allows a laboratory to easily correct a mistake that an accreditation authority could possibly make and can allow fast corrective action to occur if the authority identifies a real problem in the laboratory. Both of these reasons for a good internal quality assurance program will be more important if a national accreditation program replaces many of the multitude of current programs.

An additional reason for a good internal quality assurance system is to provide information to clients about performance on testing if a problem occurs in the laboratory. If a national accreditation program provides an assessment of a laboratory that is negative, the assessment can impact business that is not affected by the defective condition. Because data purchasers do not always understand the effect that problems have on data, it is important for the laboratories to establish control of methods so that clients can be shown the impact of problems that are found by accreditation evaluations on their data. In many cases, there is not an impact, especially if the problem occurs on a performance evaluation sample that is formulated for and analyzed by an entirely different method. In such cases, in order for the laboratory to retain its client base, it is imperative to have documentation that shows that the client's data are not affected by the problem that the accreditation authority found. If the client's data are affected by the problem, the use of past documentation to show the extent and nature of the impact of the problem on the data is even more critical. In such cases, all clients that have purchased defective data should receive an explanation of the problems inherent in the data. Only by implementing and maintaining a good quality assurance program can the impact of quality problems be minimized and can the quality of data be reconstructed.

INTERNAL QA — MORE CRITICAL THAN EVER BEFORE

It is easy to see that in cases of disputes that a laboratory contests the external appraisal of quality such as a "score" on a performance evaluation sample, that the laboratory must have evidence that the outside appraisal is not valid. The only way that a laboratory will be able to successfully refute an incorrect assessment is to provide tangible evidence that shows that the laboratory was correct. In such cases, the use of internal quality assurance records, quality control charts, the results from past assessments, and relevant current independent checks could provide the evidence necessary to indicate that the outside assessment was in error. Without such evidence provided from a comprehensive internal quality assurance program, the laboratory could not hope to cast doubt on the independent appraisal and would have to accept a possibly erroneous determination of its ability. This could be detrimental to the laboratory's image, its sales, and its reputation. Therefore, the implementation of an internal quality assurance program, even if not mandated by the accreditation body, is critical to a laboratory. In the event that a dispute arises between the accreditation program and the laboratory, a good internal QA program will prove to be even more critical than ever before.

REFERENCES

1. Bussolini, Peter, L., Alvin J. Davis, and R. Ronald Geoffrion, "A New Approach to Quality for National Research Labs," *Quality Progress*, pp. 24–27 (January 1988).
2. Worthington, J., paper presented at the Environmental Hazards Conference and Exposition, Seattle, WA (May 1990).
3. Dempsey, Carla H., paper presented at the Contract Laboratory Program Review, Fifth Annual Waste Testing and Quality Assurance Symposium, Washington, D.C., July 27, 1989.

Chapter 12

THE ROLE OF INDUSTRY IN LABORATORY ACCREDITATION AND DATA CERTIFICATION

"It costs a lot to deliver bad service".

Paraphrase of Augustine's Law XII

INTRODUCTION

Many industries have discovered that the cost of poor quality is loss of business. When supply is limited and demand is large, the quality of products is not as critical as when demand is low and supply is large. The environmental laboratory industry has enjoyed rapid growth in the past, but is now reaching a point where supply is beginning to exceed demand. When supply greatly exceeds demand, consumers become much more aware of quality and price differences. It is industry's responsibility to promote the understanding of quality differences — rather than cutting the cost and quality of their products. The time is right for the industry to appraise itself and take steps to resolve some of the problems that are impacting the industry. One such problem is isolating the laboratories that are producing shoddy products to gain an edge on competition by cutting costs. One way for industry to combat this problem is to support a uniform assessment of the quality of laboratory data.

INDUSTRY'S ROLE

This book has explained the roles of the laboratory community and the accreditation body, and also the government's role in the laboratory accreditation and data certification program. However, one of the most critical roles to address is the role of industry. Industry's role is critical since its acceptance and appropriate use of a laboratory accreditation and data certification program will either assure success or failure of the system.

First, in order for the system to be a success, it must be acceptable to data purchasers so that it helps them purchase data of the necessary quality and at reasonable prices. They must understand what both laboratory accreditation and data certification mean to them. They must understand that buying laboratory data is similar to buying any other product, and how laboratory accreditation and data certification assist them in making informed purchases.

The quality of any product must be judged based on criteria that are important to the purchaser. The purchase of data is no different from the purchase of other products. Like other products, it is possible to buy a range of different qualities of data. Laboratory data purchasers must understand that all data and all laboratories are not created equally. There are various prices of data available. There are also various types of data available, depending on what analytical process produced the data. Inexpensive data might not be a wise purchase in some cases, and in other cases it might make good business sense to purchase lower priced data. As with the purchase of any product, only the well informed purchaser understands enough about the products to tell the difference between good values and poor values.

The well informed data purchaser must understand that a laboratory that is accredited has passed a generic inspection that provides assurance that the laboratory is capable to perform certain types of analyses. The well informed data purchaser must determine if the accreditation provides any assurance of capability to perform the types of analyses that the purchaser needs. There are many different types of analyses that are performed by laboratories, and no accreditation program assesses each of these laboratory methods for all types of matrices. Since no accreditation provides in-depth assessment of capability on all methods and all matrices, only the baseline assessment aspect of accreditation and certification will likely be of interest to the purchaser. Capability of the laboratory to perform specific tests will, in many cases, need to be determined by the purchaser.

One way to determine if a laboratory is capable of performing well on special methods or matrices is to have the laboratory analyze a performance evaluation sample specific to the analysis and matrix that will be required. The purchaser should use the accreditation for all the baseline determinations, but not those which are not relevant to their specific analyses.

Likewise, data certification provides general information about the laboratory's ability to perform specified analyses. It also provides general information about a laboratory's ability to assemble appropriate data and about its internal quality assurance practices. The data for an organic analysis could provide information about laboratory contamination, accuracy, and precision — insofar as it met the required criteria — and can assist in determining if the laboratory has demonstrated the ability to meet criteria required for a special analysis. This information is useful in judging baseline laboratory capability to perform on different analyses.

The data purchaser should use the laboratory accreditation and data certification to assist in assessing a laboratory's ability to perform well on the samples to be submitted. The data purchaser should not blindly assume that the certification and accreditation will assure good work. Only if the accreditation and certification apply exactly to the purchaser's analyses would the certification and accreditation suffice. In addition, the purchaser should carefully review purchased data to determine its adequacy. This means that the purchaser or his agent should inspect the data at least on an intermittent basis, or work with the laboratory in a purchaser-supplier cooperative basis to obtain data that meets the purchaser's needs.

CRITERIA FOR DATA PURCHASE

What happens when purchasers do not use good judgement in purchasing data? In such cases, data are usually purchased according to price. The lowest bidder gets the job. If all laboratories that compete for business provide the same quality data, the purchaser is making a good decision. However, usually there is a great range in the quality of data available. Cheaper data are usually lower quality. The continual purchase of lower priced data allows the laboratories that produce poor quality data to compete with laboratories that produce quality products. Competition drives the prices of data down, but the bargain data are only a bargain if such data meets specific requirements for use. Only if the purchaser adequately inspects the data and determines if it meets the requirements will the prices of data be based on the quality of the data. Purchase of data by price with no inspection of quality tends to reinforce the production of low quality, cheap data.

Some data purchasers do not understand that good laboratory practices, including the use of internal laboratory control samples, proper storage of standard reference materials, proper storage of samples, storage of raw data, and other routine QA/QC practices, are not required of all laboratories. Changes in some regulatory methods will require good laboratory practices to be incorporated in the production of more laboratory services. However, good QA and QC are still not implemented in all laboratories. There can be a vast difference in the type and extent of laboratories' implementation of good laboratory practices and internal QA and QC procedures. There can also be a vast difference in the quality of data produced by the laboratories. Purchasing data by ordering "whatever the laboratory suggests" and at the cheapest possible price almost guarantees purchase of data that will not be useful for decision making.

Purchase of data by price alone has an adverse effect on the laboratory industry because the laboratories are rewarded only for producing low-priced data. They can either lower their expenses, including cutting QA and QC expenses, and produce data that is accordingly cheaper to produce, or they can cease to compete in the market for particular analyses. In some cases, laboratories can decide to cut corners that are not legitimate, and produce flawed data in order to make a profit. They can also legitimately lower their expenses by being more efficient or by working in a cooperative vendor-purchaser relationship with the data purchasers.

Only if data purchasers understand what constitutes shoddy data and appropriate quality data will laboratories that produce quality data be able to compete with laboratories that produce low quality data. Only when the purchasers of data understand and use laboratory accreditation and certification programs as they were designed to be used will the system work.

COST OF POOR DATA TO INDUSTRY

What is the cost of poor quality data? What will drive the purchasers of data to gain an informed position on what type of data they need for decisions? In the environmental data area, the threat of not being able to determine that property is free of environmental contamination during property transfer could be a costly mistake. Property buyers should obtain high quality data to determine if contamination exists on property before purchasing. Failure to adequately determine if contamination is present at a property before acquisition has already proven to be a very costly mistake.[1]

Acquiring good data also makes good business sense for any industry that has disposed of industrial waste at a site that is not located on the industry's premises. Suppose several industries send wastes to a waste disposal site for many years. If the waste site becomes a Superfund site, responsibility for clean up might be allocated among the industrial entities that disposed of waste at the site. Companies can be held responsible for the cost of cleaning up a site even though EPA can prove only that a tiny fraction of the site's hazardous waste barrels came from them.[2] If one industry has good historical data that supports their claims to limit responsibility for the cleanup, the data could be used to support their claims for limited liability. Poor data that cannot be defended could be useless in this situation. In such cases, the cost of potential liability is very large compared with the cost of good quality data. It is estimated that of the 8.7 billion dollars that EPA expects to be spent on Superfund in fiscal years 1987 to 1991, responsible parties will pay 3.15 billion dollars.[3]

In the future, the potential problems that poor quality data can cause will be better understood, and it is predicted that purchasers of data will use better judgement in their purchases. Some industries currently show surprisingly good judgement in their purchases, but understanding of the cost of poor quality data must be understood by more data purchasers. It is hoped that the accreditation and certification system is used as a tool by data purchasers in purchasing quality data. Such a system can help all industries make informed data purchases. The only way to promote a healthy laboratory industry that is capable and able to produce data that is the quality needed for decision making is to carefully specify the requirements for the data and then determine if the requirements have been met. Only data purchasers can perform these two steps. Therefore, only industry and government data purchasers can insure that laboratories that perform well receive business. The practice of awarding business based on price alone must be discontinued.

INDUSTRY'S RESPONSIBILITY

It is industry's responsibility to determine how much risk they can take in their purchase of data. How critical is the data in decision making? If the data are poor quality or cannot be used to support the industry during litigation, what will be the cost to the industry? Further, it is industry's responsibility to request the appropriate data rather than assuming that the laboratory can determine what data are needed. After purchasing data, the industry must determine if the data purchased meets their decision making needs. The laboratory has responsibility to produce only what was requested. The laboratory can assist in determining appropriate analyses and quantitation and detection limits if the industry provides sufficient information for them to provide this assistance. However, the industry bears responsibility for requesting data and then determining if it is usable.

Is it reasonable to hold industry responsible for determining what analyses are needed? Of course it is. A recent article on regulatory requirements by Somendu B. Majumdar, Esq.[4] is directly quoted:

> While ignorance of the law cannot be construed as a willful violation, one's lack of knowledge of the applicable laws will hardly stand as a strong defense against culpable actions. It is most imperative for all responsible parties to be up to date with all applicable laws, regulations, and amendments, if any. In the same vein, one should keep accurate records of all environmental activities, chemical releases, and wastes generated. An annual environmental audit should provide management with an understanding of how the facilities concerned are being operated. It is most appropriate that upper-level corporate management create an awareness and appreciation of various environmental laws and programs.

The article further states:

> In reality, the journey into the murky world of corporate and individual liabilities has just begun with RCRA and Superfund (and, of course, SARA). How this new part is constructed is up to Congress and the enforcement agencies. The courts, of course, will eventually decide how this path should be charted. Meanwhile, one must take all reasonable precautions to make sure that the strict guidelines of these environmental statutes are fully implemented in order to escape liability.

Only the industry knows why the data are being purchased. Is it reasonable to hold industry responsible for inspection of the product? Of course it is. The industry must determine if products meet requirements; data are simply another type of purchase that is made.

Many industries have entire processes in place for receipt inspection of products. The industries do not question their responsibility to inspect components that must meet certain specifications in order to fit into assemblies that the industry is manufacturing. Either the industry works cooperatively with the supplier or inspects a specific percentage of incoming parts before the lots are accepted. The purchase of data should also be governed by reasonable acceptance procedures. Industry has the responsibility to determine if data meets requirements and also has the responsibility to clearly request data that meets the requirements when "ordering" data from laboratories. The data purchaser is ultimately responsible for the hazardous wastes it generates and the proper processing, disposal, and analysis of the waste.[5]

TOOLS TO ASSIST INDUSTRY IN MEETING THEIR RESPONSIBILITY

In the past, the accreditation and certification schemes tended to confuse data purchasers. The proposed system of laboratory accreditation and data certification should assist data purchasers so that the system can be used to assist in their assessments of laboratories.

In addition, since the CLP has made data review a standard practice of the environmental data community, the marketplace has responded with a small industry that performs data assessments to determine data conformance with requirements. Because data review and assessment is a service that can be purchased, an industry can have data review performed, even if the expertise does not reside in the company that purchases the data. Industry must understand enough about the requirements to request an appropriate review and must have a person that is technically able to assess whether an adequate assessment has been performed. However, the actual routine assessments can be performed by third parties.

Some aspects of data assessment can be performed well by the computer. Many computer-aided data assessment software systems are currently being developed.[6,7] These software systems can easily be modified to screen large amounts of data to determine if further review is necessary and to prioritize this review. As more progress is made in the area of data useability, expert system[8] technology will become important to assist data users in making expert decisions concerning the use of data for different decisions.

In the past, the use of performance evaluation or blind samples was limited because the production of materials was limited to EPA internal program use, those that were available from the National Institute of Standards and Technology (NIST), and limited samples from chemical and standards supply houses. As was explained in Chapter 9, the availability of materials will likely increase in the future. This will allow more effective use of computer aided data review tools and better assessment of laboratory capabilities because performance can be assessed based on performance evaluation sample results. One way to reduce the cost of specific matrices for use in certain industries is to cooperate with other companies and develop a system for manufacture and distribution of these materials by trade associations or third-parties.

In the past, industry has been able to purchase data based on price alone because there were few tools available to assist them in making informed choices. This was because the tools to assess the quality of data were not well developed and were not well understood by most industrial purchasers of data. The current trend to specify requirements, and then inspect data by using the tools will gather momentum as more and better tools are developed and as industry gains an understanding of the use of the tools.

In the past, industry could easily specify the "normal analysis" for environmental analyses, which was usually the full spectrum CLP inorganic or organic set of analytes, the drinking water methods, or the EPA Office of Solid Waste SW-846 methods. Little regard was given to the actual analytes, detection limits and accuracy and precision requirements that were needed for decision making. This practice was followed by industrial and governmental purchasers of data largely because the methods and their capabilities were not well established relative to specific types of samples. The data purchaser obtained what was then the state-of-the-art because it was all that was available. The data purchaser did not have any realistic choices for alternative methods. To further complicate the situation, the type and quality of data to make decisions was necessarily determined by the quality and type of data that could be produced by the state-of-the-art methods. Therefore, the data needed for decision making and the data produced by laboratories was defined by what was technologically possible to produce at reasonable prices.

Now, many different types and qualities of data are available. The data needed for decision making can be better characterized because decision makers have models, benchmarks, health-based limits and other information that determine the quality and quantity of data needed to support decisions. In some cases, the data needed is pushing the state-of-the-art levels of quantitation even lower. In other cases, the type and quality of data needed is qualitative. The cost and quality of the different data needed for specific decisions spans the full spectrum from gross estimations needed to determine the "hot spots" in environmental investigations to extremely quantitative data that is used to determine if health-based clean-up levels have been achieved. Now that many different types of data are needed and are available, data purchasers can no longer be complacent and use whatever data has been used in the past. The cost of the different data and their suitability for decisions is vastly different. Industry must now make more informed choices in purchasing environmental data and work with laboratories to determine the best choice of data.[9]

The opportunity now exists to specify the types and quality of data needed and to purchase that data from laboratories. The present situation still promotes the use of standard CLP, SW-846, or drinking water methods because the methods are accepted by regulators and the specifications are already established by regulators. Further, the prices, QA/QC, and data deliverable requirements are already established. Industry must take the initiative to use the most appropriate method for a given purpose. Government must allow the use of less costly methods for production of data if the methods produce usable data. Industry and government might find that through better specifications of what data are needed, cost savings might be realized and more appropriate data might be produced. On the opposite end of the spectrum, industry and government might find that more costly data are needed to support some decisions. The data needed to support decisions should drive the acquisition of data. In any case, the result will be purchase of appropriate data at the best price that can be used by the purchaser for decision making.

Skeptics can point out at this juncture that data should be acquired to support decision making, but that this will never be a common practice because of the roadblocks to using data that is produced by different methods. This is a problem. However, the EPA is promoting the practice of Data Quality Objectives (DQOs), which require the decision makers to determine, a priori, what decisions must be made and then acquire the appropriate data to support the decisions.[10] The DQO process, if fully implemented, will support the use of alternative methods. In addition, as performance evaluation samples become more readily available, these materials can be used to "crosswalk" between methods and establish comparability between data sets that are acquired by use of different methods.

FUTURE RESPONSIBILITIES FOR INDUSTRY

The future holds many new responsibilities for industrial purchasers of data. The tools to support informed decision making and better data acquisition are being developed. Industry can either meet the challenge and purchase data wisely, or can utilize the last decade's technology. Industry's decision on whether to use the best data for the decision or purchase data based on price alone will have a large impact on the success or failure of a national laboratory accreditation and data certification program. The program will provide information to support industries' informed purchase of data. The tools that will allow better assessment of data will be developed to their full extent only if they are utilized by industry. Likewise, the performance evaluation materials that will serve to push the state of technology further will be encouraged only if industry purchases and uses the materials in their data acquisition and assessment procedures.

REFERENCES

1. Marcus, A. D., "S and L Bailout Faces a Costly New Complication In U.S. Hazardous-Waste Cleanup Requirement," *Wall Street Journal*, p. A16 (March 22, 1990).
2. "Supreme Court OKs Pig Farm Ruling," *Superfund*, p. 7 (February 26, 1990).
3. "NCP Pushes New Technology and Containment," *Superfund*, p. 2 (February 12, 1990).
4. Majumdar, Somendu B., "Regulatory Requirements and Hazardous Materials," *Chemical Engineering Progress* 86(5): 23, 24 (1990).
5. Lucks, John O., "Dispose Hazardous Wastes Safely," *Chemical Engineering* 97(3): 144 (1990).
6. Flynn, May J., Carla R. Schumann, and Ramon A. Olivero, *Computer-Aided Data Review and Evaluation - CADRE - Release 1.01*, prepared by Lockheed Engineering and Sciences Company, Environmental Programs, Las Vegas, NV for the U.S. Environmental Protection Agency, Environmental Monitoring and Support Laboratory, Quality Assurance Research Branch, Las Vegas, NV (1990).
7. Pandit, Nitin S., John Mateo, and William Coakley, "IQAP: An Intelligent Quality Assurance Planner for Environmental Data — Functional Requirements," presented at the American Chemical Society Expert System Conference (1989).
8. Glass, S.I., S. Bhasker, and R. E. Chapman, *Expert Systems and Emergency Management: An Annotated Bibliography*, NBS Special Publication 728, (Gaithersberg, MD: U.S. Department of Commerce, 1986) pp. 22-28.

9. Koorse, Steven J., "False Positives, Detection Limits, and Other Laboratory Imperfections: The Regulatory Implications," *Environmental Law Reporter*, pp. 10211-10222 (May 1989).

10. Data Quality Objectives for Remedial Response Activity: Development Process, U.S. Environmental Protection Agency, U.S. Government Printing Office: Washington, D.C., 54/G-87/003 (March 1987).

Chapter 13

SUMMARY OF THE PROBLEM AND THE SOLUTION

RESTATEMENT OF THE PROBLEM

> Take interest I implore you in these sacred dwellings which are designated by the expressed term, "laboratories". Demand that they be multiplied, that they be adorned; these are the temples of the future — temples of well being and of happiness. There it is that humanity grows greater, stronger and better.
>
> L. Pasteur

These "temples of the future" have reached a state of maturity and sophistication undreamed of by this early pioneer in biological experimentation. Yet the success and sophistication of technology results in mistrust, suspicion, and conflicting requirements at all levels of the laboratory industry.

As we have delineated throughout this book, the accreditation of laboratories, be they environmental, clinical, industrial, governmental, and so forth, is both a worthwhile and achievable goal. The primary hinderance to achieving this goal is the myriad of redundant and often conflicting accreditation and certification systems. We have attempted to highlight the major deficiencies in the current systems. These shortcomings all contribute to the of lack a national accreditation system that is widely accepted and highly credible. A continuation of the present approach to laboratory accreditation will ensure a still more chaotic future for the laboratory industry.

At the risk of sounding bellicose, it is apparent that the laboratory industry is suffering needless cost in both money and time in attempting to satisfy the ridiculous number of accreditation requirements. Not only does this situation cost the laboratory industry, it has a serious adverse impact on the purchasers of analytical data and the public at large. Moreover, this deplorable situation will severely curtail our nation's ability to effectively compete in the world marketplace as more uniform international trade requirements are established.

While not every ill of the laboratory industry can be attributed to the duplicative and often unnecessary accreditation requirements, a majority certainly can be. Without a national accreditation system that is widely accepted and highly credible, the present situation will continue to deteriorate.

THE PROBLEM — STATED IN A SLIGHTLY DIFFERENT WAY

Accreditation and certification programs are a direct result of the diligent decision maker attempting to determine if laboratory data are of the required quality for the decision making process. The system that one employs to determine if a product or service is of adequate quality is otherwise known as a quality assurance program. Yet, how can a good quality assurance program lead us to this present state of chaos? Have QA principles been misapplied? The classic definitions from ANSI/ASQC Standard A3-1987[1] for quality assurance and quality control are

> QUALITY ASSURANCE — All those planned or systematic actions necessary to provide adequate confidence that a product or service will satisfy given requirements for quality.
> QUALITY CONTROL — The operational techniques and activities that are used to fulfill requirements of quality.

The EPA Office of Research and Development (ORD) Quality Assurance Management Staff (QAMS)'s[2] explanation has been extracted from a document and the use of the terms is similar:

> QUALITY ASSURANCE is the process of management review and oversight at the planning, implementation, and completion stages of an environmental data collection activity to assure that the data provided by a line operation to data users are of the quality needed and claimed. Quality assurance should not be confused with QUALITY CONTROL (QC); QC includes those activities required during data collection to produce the data

quality desired and to document the quality of the collected data (e.g., sample spikes and blanks).

FY 1987 Interim Annual Report
on the Agency Quality Assurance Program
QAMS-ORD—Stan Blacker

Clearly, quality assurance, that is the ability to provide assurance that the quality of a product or service is adequate to meet the intended need, is the culmination of careful planning, implementation, and assessment to determine if the results of the planning and implementation steps have produced the desired results. In addition, quality control activities are necessary to successfully sustain quality of products and services that will satisfy given needs and provide evidence that the appropriate quality was achieved.

Somehow the real goal of the certification and accreditation systems has become obscured. Accreditation and certification systems are meant to be quality assurance programs that assist data purchasers determine if the quality of data needed is actually achieved. It appears that they have not been totally successful in achieving this goal.

SYNOPSIS OF THE SOLUTION

The proposed solution presented in this book, an independent third party approach, represents the consensus of a wide spectrum of individuals and organizations intimately involved in all aspects of the laboratory industry. As such, the independent third party approach embodies the critical attributes of acceptance and credibility necessary for the success of a national accreditation system. Further, the proposed process involves both process and product evaluation and represents a total quality approach to accrediting and certifying the products and services of various types of testing laboratories.

The proposed procedure incorporates both generic and specific criteria necessary to determine the capability and ability of the evaluated laboratories. Concomitantly, the specifications and criteria are sufficiently flexible to allow an accreditation and certification system to be designed to meet the requirements of any organization requesting a laboratory accreditation evaluation. By utilizing a third party, the proposed procedure encompasses the critical requirements of maintaining technical competence in all areas; being independent while maintaining accountability to the organizations requesting the evaluation; ensuring and maintaining credibility; and ensuring widespread and improved acceptance in all organizations involved with or using the services of the laboratory industry.

POTENTIAL FUTURE APPLICATION OF THE SOLUTION

The independent third party approach to accreditation and certification is capable of being applied in various situations. The implementation of this approach in the many and diverse types of testing laboratories has been discussed. The proposed approach is capable of being implemented in a tiered fashion and has the flexibility to be designed for any organization that produces data (or more generally, products and services). Any situation — technical, research, manufacturing, and so forth — for which consensus standards can be developed will benefit from adaptations of this proposed approach to accreditation and certification.

LIMITED RECOMMENDATIONS

Caveat emptor — let the buyer beware — can no longer be the watch word of either purchasers or producers of laboratory products and services. The complexity of our technologically based society and our wide-ranging dependence on analytical data require us to have confidence in the products and services of the laboratory industry. Consequently, the time is long past for the development and implementation of a national system for laboratory accreditation and data certification. Both assessment of laboratory capability and monitoring of data quality are necessary for the system to be credible. A system that does not include both will not have value to data purchasers. The authors have presented a procedure for instituting a credible system that includes capability assessment and monitoring of data quality. It is our fervent hope that logic and reason, rather than self-interest and short-sighted gain, prevail. If a credible system is implemented, we can all echo the words of Louis Pasteur — laboratories will indeed be the "temples of the future". A productive and stimulating future for the laboratory industry, the purchasers of laboratory goods and services and the public as a whole will be the end result.

REFERENCES

1. "Quality Systems Terminology," ANSI/ASQC Standard A3-1987 (Milwaukee, WI: American Society for Quality Control, 1987).
2. Blacker, Stanley, "FY 1987 Interim Annual Report on the Agency Quality Assurance Program," U.S. Environmental Protection Agency, Office of Research and Development, Quality Assurance Management Staff, Washington, D.C.

Chapter 14

THE FUTURE IN THE 90's

"Everyone says something must be done — but this time it looks like it might be us."

Will Rogers

INTRODUCTION

If the state of affairs is to be improved in the future, all people that can promote the improvement of data must join together in the effort. This book was written to offer a possible solution and provide structure for improvement based on all affected parties cooperating in the design of the final system.

Before we predict what might happen in the future, and then further project our hopes for the future, let us reflect on the state of affairs as they exist now and how this affects the design of the system.

THE CLP — IS IT A GOOD MODEL — OR A FAILURE?

Some readers might legitimately question patterning a national system after the CLP. This is a real concern, because as these words are written, a significant number of laboratories that hold CLP contracts are under investigation for fraud, waste and abuse by the Office of the Inspector General.[1] Does this mean that the CLP cannot be used as a pattern for a national accreditation and data certification program? The answer to this question is an emphatic NO! The investigations will show that the program has many problems, but not that the design is incorrect. We must now address the weaknesses in the system that the investigations are exposing, and then design the system so that the effects of these weaknesses are minimized in the national system.

It is important for critics of the CLP to realize that the reason that problems can be found and assessed is because the CLP requirements are so carefully detailed that deviations from these requirements can be proven. Also, because written records are required, collected, assessed, and maintained, written evidence of deviations can be found. If it were not for these requirements, evidence necessary to prove departure from adherence to requirements would be unobtainable.

Some people might think that because CLP laboratories are being investigated, they are poorer performers than other laboratories. Consider the fallacy of that reasoning. If one were to cheat on requirements, it would not be wise to cheat on the CLP program, because it is the only program that can easily prove that deviations from requirements have occurred. In many other programs, such as those used by industry and other accreditation systems, the proof of deviation from requirements is nearly impossible to obtain. The critics of the CLP surely do not think that the laboratories would cheat on the only system that is capable of proving deviations. It is not logical to assume that all other systems are flawless simply because there is no way to prove that problems exist. It would be more reasonable to assume that a greater magnitude of cheating exists in programs in which deviations cannot be easily discovered.

Why has cheating occurred in the CLP? One cause might be that laboratories cheat to make a greater profit.[2] This might be the case in isolated instances, but there is probably a better explanation of the apparent widespread incidence of the problem. As has been mentioned previously, the CLP is currently considered the de facto laboratory certification program for environmental analyses. Because it is considered to be the only credible system by many data purchasers, many purchasers acquire data only from laboratories that have a CLP contract. In order to compete for this business, laboratories must be a part of the CLP. The "membership" in the CLP cannot be acquired by an inspection, payment of a fee, voluntary submission of data for assessment, or other usual mode of "membership" in an accreditation or certification program. To gain a CLP contract, laboratories must successfully bid on government contracts. As was explained in Chapter 5, this means that the successful bidders must meet technical requirements and be in the competitive cost range of the bidding. Effectively, this means that laboratories must bid low prices to assure acceptance of the bid by the government. The effective bid price of many laboratories may be lower than the cost that the laboratory expends to perform the analyses. The laboratories may take this loss in order to become "members" of the CLP. They need to be a part of the CLP to be credible to other purchasers of environmental data.

Since the CLP "membership" is required for laboratories to effectively compete in the environmental marketplace, and the prices bid to the government are not reasonable, the impetus in the laboratories might well be to cut quality control, cut corners, and perform only the minimum-acceptable work on the government contract samples. The fine line between cutting-corners and waste, fraud, and abuse, might now have been breached by some laboratories. Lack of a credible, national environmental laboratory accreditation and data certification program that can be utilized by laboratories to indicate competence has contributed to the misuse of the CLP and the problems that are now evident in the program. Some blame the administrators of the CLP for the present state of affairs. In the spirit of looking to the future, it is not constructive to establish blame. The point is to use this experience to learn from and assist in designing a better system. The CLP has sufficient monitoring information on the laboratories so that the information can be scrutinized now to determine if signs of problems were evident, but were unrecognized. This information will assist in designing better ways to monitor laboratories in the future.

Implementation of a credible, national system that encompasses more than the CLP analyses will serve to minimize the impact of the CLP. The problems that now occur in the CLP because of low bid prices and because such a large number of laboratories must be monitored for performance might be alleviated by a national system. In addition, since the fundamental principles governing the national program will have their roots in the CLP, it is not unreasonable to expect that government programs such as Superfund will use the program, insofar as it is possible to do so.

PROJECTIONS FOR THE FUTURE FOR ENVIRONMENTAL DATA

The future holds many changes in the way data will be acquired and used. Based on the progress that has been made in the short history of environmental measurements, it is realistic to project many improvements that might occur in the next few years. The improvements will be facilitated by the different ways of determining what constitutes acceptable data for use in regulatory programs that is rapidly emerging. The concept of performance-based requirements is supported by many people within EPA.[3,4] Performance-based requirements are different than methodology-based requirements because the former states what acceptable performance characteristics of the data are versus the latter's delineation of methods that must be used to produce data. The use of performance-based requirements allows much greater latitude for using innovative technology. The only requirement is that the data produced using an alternative method meets performance requirements. Specifying only method-based requirements limits production of data to specific methodology that might not produce the most effective and best data. Acceptance of performance-based requirements will allow more rapid advancement of new technologies and will improve future environmental chemistry capabilities.

As has been stated in several places in this book, the use of performance evaluation materials will facilitate the direct comparison of data sets produced by different methods. The proper use of the materials will allow better assessment of laboratory data and will allow laboratory assessment to identify areas of laboratory problems. As the use of these materials becomes more common, the manufacture of the materials will be better understood and the types of materials available will increase. As routine production procedures for the materials become fully developed, the cost of the materials should become lower. These materials could greatly simplify data assessment and method equivalency determinations. Instrument manufacturers could find it useful, in the future, to manufacture or distribute performance evaluation materials to support their innovative technologies. Such supporting materials could accelerate acceptance of the technologies because data equivalence and technology assessment could be facilitated by using the materials.

In the past, the term "litigation quality" data has been used to describe specific documentation and chain-of-custody procedures. In the future, sophisticated decision makers and data purchasers will take a more logical approach in determining what constitutes "litigation quality" data. "Litigation quality" might be redefined to state that any data that is acquired must have minimal chain-of-custody and documentation associated with it. The number reported from a sample must be shown to be associated with the sample and to the location from which the sample was taken. Without this minimal chain-of-custody documented, a number cannot legitimately be used for any decision. The concept of "litigation quality" data must be addressed, and the minimum requirements for acceptability of data must be delineated so that it is obtained for all data. The National Enforcement Investigations Center (NEIC) has made progress in defining the important aspects of data that facilitates its use as evidence in enforcement proceedings for Superfund.[5] Their efforts should be applauded and promoted by data users so that more types of data can be effectively used for decision making and in subsequent litigation, if necessary.

In the past, the standard format for data has become the CLP "diskette deliverable." This has occurred by default. That is, in the absence of another format for data array, this format has been used by most laboratories. Unfortunately, this format provides a standard data delivery format for only a few types of data — largely standard CLP organic and inorganic data. In order for data to be accepted by many different clients and for effective assessment of data to be accomplished, a standard delivery format or a way to handle several formats for data needs to be developed. Until minimum requirements for data deliverables are established for all data, it will be impossible to design a standard data delivery format. Until minimum requirements for data delivery have been established, it will be impossible to specify the minimum assessment of the data to determine if it is suitable for use in immediate and subsequent decisions. Further, it is impossible to use data from many different sources in a common database unless a common format is specified. Data cannot be effectively used for making global decisions unless transfer and access to information is facilitated by data format and reporting standardization. This problem has been acknowledged for many years, but the minimum requirements for data reporting have not yet been fully established. Hopefully, the future will bring consensus on this issue and a standard format for data will emerge.

In conjunction with a standard format for delivery of data, a more standardized approach to large database storage and retrieval must be established. This is a large and formidable task, because in the absence of minimum data delivery requirements, the structure and content of the database cannot be well defined. In addition, the issues of data security, data quality and integrity, and data reliability and defensibility are issues that must be adequately addressed before automated data technology can be fully utilized.[6]

The barriers to transfer of data across government programs and across media will hopefully disappear for several reasons. First, an increased focus in the state of the total environment that is not media specific and not government agency specific will emerge. This will occur because environmental problems are rapidly being recognized as global problems. As transfer of data between programs and media becomes easier due to use of generic format data deliverables and "performance-based" methodology, information will be used for many interrelated purposes among government programs and agencies. The information collected in programs such as RCRA permitting, waste water discharge permitting, drinking water testing, air-emissions discharge, Superfund site assessment and long-term monitoring will all be used to determine the condition of the environment rather than being used for only one purpose. In the future, it is hoped that all data that is collected for specific decisions will be used to the largest extent possible — in order to gain a better assessment of the global state and in order that future environmental solutions can become more effective.

In keeping with the hope for the future that environmental data will be utilized effectively, it is also hoped that biological data for environmental assessments (biomonitoring) will be used to augment and replace, where possible, analyte by analyte physical chemistry parameter analyses. The use of biomonitoring data to enhance physical and chemical data provides information about real impacts to biota, provides an effective tool for use in trend assessment, and provides an integrated risk-based approach to assessments. Biomonitoring to determine what kinds and levels of contaminants cause adverse effects to human health and the ecology could prove to be a better way to determine what contaminants are most detrimental to people and ecology. Biomonitoring is already being used, to a limited degree in integrating water pollution control programs.[7] Hopefully, this trend[8] will be expanded in the future to encompass all types of media to assure an integrated, global, biological and chemical approach to environmental assessment and monitoring.

FINAL HOPE FOR THE FUTURE

The final hope for the future is for decisions to be based on the quality of data needed for each decision and that all data will be effectively utilized to extract the best technical information possible from it. In order for this to be achieved, data reporting must become more standardized and data quality must be described in the reports, assessed by the purchaser of data, and recorded with the data when it is entered into databases.

In order for this hope to be realized, the tools to facilitate the acquisition and assessment of data for decision making that are currently being developed must be utilized by laboratories. It is hoped that industry and government will work cooperatively with laboratories to obtain the most suitable and cost-effective data for specific purposes. It is also hoped that cooperative ventures will extend technology further to enhance the types of information that will be available for use in the coming years.

Finally, it is hoped that this laboratory accreditation and data certification program — the System for Success — can be implemented to encourage minimum acceptable standards across the environmental laboratory community. The System cannot be designed only to inspect capability to perform, because such systems provide a false sense of assurance to data purchasers that laboratories are competent to produce useable data. The System must be designed to include both capability and performance assessments, or it will do more harm than good for data purchasers. Further, the System must be designed to incorporate alternative methods such as bioassay techniques, new concepts such as "performance based" requirements and all other ideas that were suggested as hopes for the future so that it does not become obsolete. It is hoped that the System will make it easier to acquire good data, facilitate the assessment of data to determine its suitability for use, and will promote data utilization for global decision making and monitoring efforts.

REFERENCES

1. Himelstein, Linda, "Superfund Effort Jeopardized by Suspect Data," *Legal Times* (April 23, 1990).
2. Felsen, Harvey G., "What's Wrong With Our Industry?" *Environmental Testing Advocate* II(1): 3 (1990).
3. Fairless, Billy J., and Dale I. Bates, "Estimating the Quality of Environmental Data," *Pollution Engineering*, pp. 108–111 (March 1989).
4. "Instrumentation '90" *Chemical and Engineering News* 68(23): 26–32 (1990).
5. Smith, Paula and D. Roche, paper presented at the U.S. EPA Quality Assurance Caucus, Denver, CO, (May 1989).
6. "Automated Laboratory Standards: Current Automated Laboratory Data Management Practices — DRAFT," prepared by Computer Sciences Corporation (CSC), Research Triangle Park, NC, for U.S. Environmental Protection Agency, Office of Information Resources Management, Research Triangle Park, N.C. (1989).

7. "Bios Field Survey Component Manager's Guide," U.S. Environmental Protection Agency, Office of Information Resources Management, (November 1989).

8. "Availability, Adequacy, and Comparability of Testing Procedures for the Analysis of Pollutants Established Under Section 304(h) of the Federal Water Pollution Control Act, Report to the Committee on Public Works and Transportation of the House of Representatives, EPA/600/9-87/030, pp. 1.1–1.2 (1988).

INDEX